Lecture Notes in Control and Information Sciences 207

Editor: M. Thoma

Springer-Verlag London Ltd.

R. Gabasov, F.M. Kirillova and S.V. Prischepova

Optimal Feedback Control

Springer

Series Advisory Board

Authors

Rafail Gabasov, MS, PhD, Doctor of Science

Department of Applied Mathematics and Informatics,
Byelorussian State University, F. Skorina Av. 4, Minsk 220080, Belarus

Faima M. Kirillova, MS, PhD, Doctor of Science
Svetlana V. Prischepova, MS, PhD

Institute of Mathematics, Academy of Sciences, Surganov str. 11,
Minsk 220072, Belarus

ISBN 978-3-540-19991-5 ISBN 978-3-540-39381-8 (eBook)
DOI 10.1007/978-3-540-39381-8

British Library Cataloguing in Publication Data
A catalogue record for this book is available from the British Library

Typesetting: Camera ready by authors

69/3830-543210 Printed on acid-free paper

CONTENTS

INTRODUCTION

This book describes a new approach to constructing optimal control of the feedback type and algorithms for the synthesis of optimal systems in terms of the new approach

The optimal systems synthesis problem is central to control theory. It arose as a natural development of the classical theory of regulation in the early 1950s. Some eminent specialists have studied the synthesis problem (J. Ackermann [1], R. Bellman [4], V.G. Boltyanski [5], S.S.L. Chang [8], J. Cruz [10], C.A. Desoer [7,23], A. Feldbaum [13,14], R. Kalman [24], H.W. Knobloch [25], H. Kwakernaak [26], E.B. Lee [32], A.M. Lyotov [34], M. Thoma [41], M. Vidyasagar [42]).

As is known in the theory of regulation, there are two principles of control: program control and feedback control. Two types of systems correspond to these: open-loop and closed-loop systems.

For a long time the main efforts of the specialists in automatic regulation have concentrated on methods of construction of controllers, which being included in feedback, provide processes with definite properties.

The problem of constructing feedback which appeared in the context of the theory of optimal control in the 1950s turned out to be extremely difficult. It is therefore not accidental that the principal success in mathematical theory of optimal control arose from investigating program controls this is famous Pontryagin's maximum principle [38]. Generally speaking, attempts to solve the optimal synthesis problem with the help of the maximum principle and the second fundamental approach of optimal control theory - Bellman dynamic programming [4], have failed. Some progress was achieved in one very particularly actual for applications area of optimization, namely in the field of linear systems with quadratic criteria, without account of any constraints on control and system state (R. Kalman [24], A.M. Lyotov [34]).

We consider the linear terminal problem

$$I(u) = c'x(t) \longrightarrow max, \qquad (1)$$

$$\dot{x} = Ax + bu, \; x(0) = x_0, \qquad (2)$$

$$x(t^*) \in X^* = \{ x \in \mathbb{R}^n : Hx = g \}, \tag{3}$$

$$| u(t) | \le 1, \; t \in T = [0,t^*] \tag{4}$$

where $x \in \mathbb{R}^n$, $u \in \mathbb{R}$, $g \in \mathbb{R}^n$, $rank\ H = m < n$.

This problem is the simplest one since only linear functions in the linear equation (2) and linear relations are present in its formulation. But problem (1)-(4) contains all principal elements of the general optimal control problem. This problem becomes trivial if even one element is removed. The synthesis problem has not been solved for it yet. It is considered to be initial problem in our approach since it is convenient for stating the main elements of the approach. As is known, an optimal program control is any piecewise continuous function which satisfies the constraints (4) and generates an optimal trajectory $x^o(t)$, $t \in T$, of system (2), getting into the terminal set (3) at t^* and maximumizing the quality criterion (1).

We understand an optimal control of the feedback type to be a piecewise continious function $u^o(x,t)$, $x \in \mathbb{R}^n$, which for every moment $\tau \in T$ and every initial state x_o from the set of controllability $S(\tau)$ generates the trajectory of the system

$$\dot{x} = Ax + bu^o(x,t), \; x(\tau) = x_o$$

coinciding with the optimal trajectory $x^o(t)$, $t \in T$, of problem (1)-(4).

The main reason why engineers prefer control of the feedback type to program control is that real movements of dynamic systems are described not by the equation

$$\dot{x} = Ax + bu, \; x(0) = x_o,$$

but by

$$\dot{x} = Ax + bu + w(t), \; x(0) = x_o, \tag{5}$$

where $w(t)$, $t \in T$, is an unknown n-vector function. Under the conditions (5), program control cannot provide even admissibility of trajectories. Feedback control can cope successfully with a great number of perturbations.

Before stating the new approach to the synthesis problem we shall analyse the classical statement.

1. The program control $u^o(t)$, $t \in T$, is often criticized because it is constructed before the system is activated and is not changed after it is switched on. However, a similar criticism is valid for feedback control $u^o(x,t)$, $x \in \mathbb{R}^n$, $t \in T$, because the last function must be constructed before the initial moment $t = 0$ and its value changes in the process only with respect to the change $x^o(t)$, $t \in T$.

Thus in constructing feedback it is necessary to do much preparatory work on computation and tabulation of functions of $(n+1)$ variables. In practice the purpose of the controller is reduced to indication of the value $u^o(x,t)$ which corresponds to the current variables x, t. It is clear that only in rare cases (as indicated above in connection with the papers of R. Kalman and A.M. Lyotov) may we rely on finding on a formula, an explicit expression for $u^o(x,t)$.

In the general case we inevitably come across with the so-called "curse of dimension". This phenomenon was noted by R. Bellman in dynamic programming.

2. Controls of the feedback type are constructed according to the presence of unknown perturbations (5) in the real processes. But the perturbations themselves are not taken into account in the classical statement of the synthesis problem. So far, it is clear that the type of optimal controller and its properties will depend on the accepted model of perturbations.

At present, three models of perturbations are widely used: 1) passive stochastic; 2) active game type; 3) passive given by sets. The first model is investigated in terms of optimal stochastic control [2,9,31,33,39] and the second one in the theory of differential games [29]. Lately much attention has also been paid to the third model of perturbation [27-30,40].

3. A function $u^o(x,t)$, $x \in \mathbb{R}^n$, $t \in T$, providing optimal

feedback in many problems (and, in particular, in problem (1)-(4)) is discontinuous in x. So equation (5) is not already the classical object of differential equations and because of this, difficulties with the existence and uniqueness of solution to (5) and also with the correct statement of the optimal synthesis problem may arise.

4. The classical statement of the optimal feedback problem is static and does not cover control by a concrete process under conditions of unknown perturbations. Such situations are considered as if all functions $u^o(x,t)$ are always and at once necessary. Actually, in the real control processes only the values $u^o(x,t)$, $t \in T$, along the trajectories are necessary.

In other words in the classical statement of the synthesis problem the real process and velocity boundedness of real time are not taken into consideration. However, in real life we constantly meet these circumstances. Although,under conditions of continuous work over a finite time we can fulfil a sufficient amount of work, for each small interval the volume of work fulfilled may be small. The controller may operate in the same way in the process of system action. It works to treat perturbations in real time, the integral action of which to a system may be large. The capacity of the class of perturbations depends on the speed of the controller.

Controllers constructed with the indicated properties may naturally be called dynamic.

5. The classical statement of the synthesis problem is closely connected with very old notion of solution of mathematical problems which is traditionally taught in schools: there is an obvious wish to obtain a formula for solution of a complex problem. The desire of engineers to make a box (Fig.1) true for any case, corresponds to this mentioned situation.

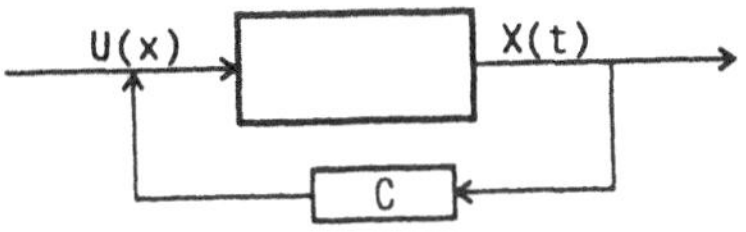

Fig. 1

A controller acts according to some formula. As is known, mathematicians rejected long ago the similar notion of explicit solution because of the impossibility of reaching this aim in many cases (recall the classical problems of solution of algebraic and differential equations; only in rare cases can solutions in radicals and quadratures be obtained). Numerical algorithms for constructing approximate solutions which allow us to obtain exact answers for any concrete set of parameters came to replace the explicit-formula solutions. The appearance of computers promoted the success of a new notion of solution. Such a notion of solution became very convenient for constructing optimal program control. In our view the problem of synthesis of optimal systems contains one more notion of solution of mathematical problems. It is closely connected with the specific character of the control process in real time. It seems that this has been not investigated systematically in mathematics. We talk about continuously forming extremal problems in real timeand the corresponding continuous correction of current solutions of these extremal problems. Each of the problems is of a program type and needs considerable computer time for its solution. However, since the parameters of continuous series of extremal problems are connected in a continuous way, it is advisable to consider them not as independent but as elements of a unique continuous chain of problems. In the modern theory of extremal problems there exist methods of correction of solutions and they are much more effective measured in computer time than methods of complete solution. Consequently, the notion of solution of the totality of extremal problems as a process of continuous successive correction of current solutions is natural for the theory of optimal feedback control.

We can interpret the proposed approach as a unification of ideas of optimal program control and invariant embedding. As mentioned above, program control is the main object of the maximum principle and invariant embedding is the main means of dynamic programming. These methods were natural for linear feedback control and corresponded to the level of technology of the 1940s and early 1950s. During the last 40 years the role of computers in engineering has greatly increased. It has made a rapid leap during this time.

In our approach the problem is embedded in one parametric family of problems along a realizing trajectory. Real time is a parameter of the family.

6. It should be stressed that the solution of the classical synthesis problem is special. In the 1950s it was natural to use results of the classical theory of regulation. It represented feedback in the form of hard schemes containing mechanical, electrical, magnetic, acoustic and hydraulic elements (links). From the modern point of view such feedback can be called hard. As mentioned above, computer engineering made a rapid leap which has been called the modern scientific and technological revolution. Among the achievements in computer engineering, microprocessors occupy a special place. In designing an algorithm of microprocessor's operation we can realize very complex feedback. We can change propeties (characteristics) of feedback by changing only programs. It is natural to call this type of feedback flexible. Flexible feedback will be studied in this book.

Let us sum up. What does the proposed approach give to the problem of synthesis of optimal systems give? If there is no need to parry sliding motion then controls generated by the controller $u^*(x,t)$ and the trajectory $x^*(t)$ coincide with the optimal program control and the corresponding trajectory (if the chosen loop system is optimal in the sense of the classical statement). If a sliding regime is possible then the controller will work and the classical statement of synthesis must be specified.

The approach to solving the synthesis problem is based on schemes worked out in Minsk during 1975-1990 (R. Gabasov, F.M. Kirillova et al.[15-22]).

The main object of study is discrete systems obtained from continuous ones (excepting Chapter 2).

In Chapter 1 an optimization method of discrete control systems is proposed. Its description is necessary for understanding the following chapters.

In Chapter 2 some questions connected with optimization of continuous control systems in real time are considered.

In Chapter 3 the problem of synthesis for control systems under conditions of uncertainty is considered. Different controllers for systems operating under various information conditions are constructed.

In Chapter 4 the approach proposed in Chapter 3 is generalized. Optimal feedback control is designed on the basis of incomplete and inexact data provided by the measuring device.

In the Appendix the adaptive method of linear programming is presented. This method was described by R. Gabasov, F.M. Kirillova and O.I. Kostyukova in the late 1980s. Its constructions are used in Chapter 1 to justify a finite algorithm to solve the terminal optimal control problem under restrictions on terminal states and control.

Sections 2.4 and 4.2, 4.3 were written by the authors jointly with Dr. P.V. Gaishun. We wish to thank him.

We wish to extend also our thanks to Dr. A.S. Chernushevich for the translation of Appendix and to Dr. O. Kunchev for editorial assistance on the book.

CHAPTER 1

OPTIMIZATION OF LINEAR SYSTEMS

1.1. DISCRETE CONTROL SYSTEMS. THE CAUCHY FORMULA.

We consider the discrete control system

$$x(t+1) = A(t)x(t) + B(t)u(t),$$
$$x(0) = x_0, \; T = \{ 0, 1, 2, \ldots, t^* \} \tag{1.1}$$

where $x(t) = (x_1(t), \ldots, x_n(t))$ is a state of the system at the moment t; $u(t) = (u_1(t), \ldots, u_r(t))$ is a control vector, $A(t)$, $B(t)$, $t \in T$, are $n \times r$ matrix functions characterizing system properties and x_0 is an initial state.

Each control $u = (u(t), t \in T)$ by virtue of (1.1) stays in correspondence with the only sequence $x = (x(t), t \in T \cup t^*)$ which is called the discrete system trajectory. In the space of variables $x = (x_1, \ldots, x_n)$ (state space or phase space of the discrete system) discrete system trajectories are presented by broken lines passing through the states $x(0)$, $x(1)$, $\ldots$, $x(t^*)$.

To calculate the states $x(t)$ of the discrete system at any moment of time t by the known initial state x_0 and the given control $(u(\tau), \tau = 0, 1, \ldots, t-1)$ we shall derive a formula for solving (1.1).

The trajectory of the system (1.1) corresponding to the control $u(\tau)$, $\tau = 0, 1, \ldots, t-1$, is determined by

$$x(\tau+1) = A(\tau)x(\tau) + B(\tau)u(\tau), \; \tau = 0, 1, \ldots, t-1. \tag{1.2}$$

Let us multiply both parts of the identity (1.2) by the $n \times n$ - matrix function $F(t,\tau)$ and sum over τ from 0 to $t-1$:

$$\sum_{\tau=0}^{t-1} F(t,\tau)x(\tau+1) = \sum_{\tau=0}^{t-1} F(t,\tau)A(\tau)x(\tau) + \\ + \sum_{\tau=0}^{t-1} F(t,\tau)B(\tau)u(\tau). \tag{1.3}$$

Since

$$\sum_{\tau=0}^{t-1} F(t,\tau)x(\tau+1) = F(t,t-1)x(t) + \sum_{s=0}^{t-1} F(t,s-1)x(s) - \\ - F(t,-1)x(0) \tag{1.4}$$

then assuming

$$F(t,t-1)=E \tag{1.5}$$

and substituting (1.4) into (1.3) we get

$$x(t)= F(t,-1)x(0)+ \sum_{\tau=0}^{t-1}(F(t,\tau)A(\tau)-F(t,\tau-1))x(\tau)+ \tag{1.6}$$

$$+ \sum_{\tau=0}^{t-1}F(t,\tau)B(\tau)u(\tau).$$

Let $F(t,\tau)$, $\tau = 0,1,\dots,t-1$, be a solution of the equation

$$F(t,\tau-1) = F(t,\tau)A(\tau), \ \tau = 0,1,\dots,t-1. \tag{1.7}$$

From (1.5) and (1.7) the function $F(t,\tau), \tau = 0, 1, \dots, t-1$, is calculated uniquely.

With regard to (1.7) the expression (1.6) will be reduced to

$$x(t)= F(t,-1)x(0) + \sum_{\tau=0}^{t-1} F(t,\tau)B(\tau)u(\tau). \tag{1.8}$$

The relation (1.8) between $x(t)$ and x_0, $u(\tau)$, $\tau = 0, 1, \ldots, t-1$, is called the Cauchy formula. The matrix function $F(t,\tau)$ given at $\tau = 0, 1, \ldots, t-1$, by (1.7) with the initial condition (1.5) is said to be the fundamental matrix of solutions of the system

$$x(t+1) = A(t)x(t).$$

In applied problems it is often not the states of discrete systems are of prime interest but their output signals connected to $x(t)$ by the equality

$$y(t) = H(t)x(t) \tag{1.9}$$

where $H(t)$, $t \in T \cup t^*$, is an $m \times n$ matrix function of the output device parameters.

Definition 1.1. The discrete system (1.1) is called controlled on T with respect to the output (1.9) if for any m vector g, there is a $u(t)$, $t \in T$, such that the corresponding output signal at t^* takes values g, i.e.

$$y(t^*)=H(t^*)x(t^*)=g \tag{1.10}$$

We assume that

$$\mathrm{rank}\ H(t^*) = m \le n.$$

Write the equality (1.10) in terms of the Cauchy formula (1.8):

$$\sum_{\tau=0}^{t^*-1} H(t^*)F(t^*,\tau)B(\tau)u(\tau) = g - H(t^*)F(t^*,-1)x(0). \tag{1.11}$$

For solvability at any $g \in R^m$ of the linear equation (1.11) relative to $u(t)$, $t \in T$, it is necessary and sufficient that

$$\mathrm{rank}\ (\ H(t^*)F(t^*,\tau)B(\tau),\ t \in T\ ) = m. \tag{1.12}$$

Theorem 1.1. For controllability of the discrete system it is necessary and sufficient that the condition (1.12) be fulfilled.

Corollary 1.1. Suppose that $r=1$, $B(t) = b(t)$. In order that the discrete system (1.1),(1.9), be controllable, it is necessary and sufficient for there to exist a set $T_{sup} = \{ t_1, t_2, \dots, t_m \}$ of $t_j \in T$, $j = \overline{1,m}$, such that the $m\times m$ matrix

$$P = (h(t),\ t \in T_{sup})$$

drawn up from the columns

$$h(t)= H(t^*)F(t^*,t)B(t),\ t\in T_{sup}$$

be non-singular.

1.1.1. Terminal optimal control problem.

Let $u(t), t\in T$, satisfy the inequality

$$f_*(t) \le u(t) \le f^*(t),\ t \in T \tag{1.13}$$

where $f_*(t)$, $f^*(t)$, $t \in T$, are given r-vector functions.

Definition 1.2. A control $u(t), t\in T$, and a corresponding trajectory $x(t)$, $t \in T \cup t^*$, are called admissible if, for the given m-vector g and the $m\times n$ matrix H, $rank\ H = m$, we have

$$Hx(t^*)=g.$$

Geometrically this means that admissible controls generate trajectories with the initial state x_0 reaching the plane $Hx = g$ at $t=t^*$.

We shall estimate admissible controls with the help of the quality criterion

$$J(u) = c'x(t^*) \tag{1.14}$$

defined by an n-vector c on the final (terminal) states of the discrete system.

Definition 1.3. We shall call the admissible control $u^o(t)$, $t \in T$, and the corresponding trajectory $x^o(t)$, $t \in T \cup t^*$, optimal if upon them the criterion (1.14) attains the maximum value

$$J(u^o) = \max J(u).$$

The terminal problem of optimal control consisting of the construction of $u^o(t)$, $t \in T$, can be written in the compact form

$$J(u) = c'x(t^*) \longrightarrow \max, \; x(t+1) = A(t)x(t) + B(t)u(t),$$

$$x(0) = x_0, \; Hx(t^*) = g, \tag{1.15}$$

$$f_*(t) \le u(t) \le f^*(t), \; t \in T = \{0, 1, 2, \dots, t^*\}.$$

Later on, side by side with optimal controls we shall look for ε-optimal control of problem (1.15).

Definition 1.4. An admissible control $u^\varepsilon(t)$, $t \in T$, and a corresponding trajectory $x^\varepsilon(t)$, $t \in T \cup t^*$, is called suboptimal (ε-optimal) if

$$J(u^o) - J(u^\varepsilon) = c'x^o(t^*) - c'x^\varepsilon(t^*) \le \varepsilon.$$

1.1.2. Optimal control problem as a linear programming problem.

Problem (1.15) at t^*-1, when the number of steps (process duration) equals one, turns into a static problem and takes the form

$$c'x(1) \longrightarrow \max, \; x(1) = A(0)x_0 + B(0)u(0), \tag{1.16}$$

$$f_*(0) \le u(0) \le f^*(0), \; Hx(1) = g$$

which coincides with the canonical form of the linear programming problem

$$\bar{c}'z \longrightarrow max, \quad \bar{A}z = \bar{b}, \; d_* \le z \le d^*$$

where we have put

$$z = u(0), \; \bar{c} = c'B(0), \; \bar{A} = HB(0),$$

$$\bar{b} = g - HA(0)x_0, \; d_* = f_*(0), \; d^* = f^*(0).$$

On the other hand the dynamic problem (1.15) may be interpreted as a particular case of the linear programming problem. In fact, introduce the variables $u(0), u(1), \ldots, u(t^*-1), x(0), x(1), \ldots, x(t^*)$. Their total number equals $rt^*+n(t^*+1)$. Linear equality constraints

$$x(t+1) = A(t)x(t) + B(t)u(t), \; x(0)=x_0, \; t \in T, \; Hx(t^*)=g$$

are imposed on these variables. Their total number equals $n(t^*+1)+m$. Furthermore the problem (1.15) has $2t^*r$ linear inequality constraints

$$f_*(t) \le u(t) \le f^*(t), \; t \in T.$$

If we still take into account that in problem (1.15) the function $c'x(t^*)$ is linear and it is maximized, then problem (1.15) may be interpreted as a static linear programming problem of size $[\, n(t^*+1)+m] \times [rt^* + n(t^*+1)]$.

In applied problems the values n, m, r are not often large ($r \le 3$, $m < n \le 10$), but t^* (the number of steps) is, as a rule ($t^* =$ 100 to 1000). For large t^* the size of the equivalent static problem of linear programming appears to be large ($10^4 \times 10^4$). Apart from large size, problem (1.15) being interpreted as a static linear programming problem has a number of peculiarities which influence negatively the effectiveness of general methods for solution of linear programming problems.

However, it should also be noted that problem (1.15) has one peculiarity which is effectively used in modern realizations of the simplex method. This peculiarity consists of strong rarefaction of the expenditure matrix of the equivalent linear programming problem, i.e. this matrix does not have a large number of non-zero elements. This positive feature of problem (1.15) only softens the indicated negative features. Therefore the solution of problem (1.15) directly by general linear programming methods is not reasonable due to their low efficiency in this specific case.

There exists another method of solving problem (1.15) by using general linear programming methods. It is based on taking into account the specific structure of constraints - equalities defined by the system dynamics.

With the help of Cauchy's formula we eliminate the variables $x(0), x(1), \ldots, x(t^*)$ from problem (1.15). The equation

$$x(t+1) = A(t)x(t) + B(t)u(t),\ x(0) = x_0,\ t \in T$$

is replaced by the equality

$$x(t^*) = F(t^*,-1)x(0) + \sum_{t=0}^{t^*-1} F(t^*,\tau)B(\tau)u(\tau)$$

and it is substituted into the criterion and the terminal constraints. We get a static linear programming problem equivalent to the dynamic problem (1.15):

$$\sum_{\tau=0}^{t^*-1} c'F(t^*,t)B(t)u(t) \longrightarrow max,\ f_*(t) \le u(t) \le f^*(t),\ t \in T,$$

$$\sum_{\tau=0}^{t^*-1} H(t^*)F(t^*,t)B(t)\ (t) = g - H(t^*)F(t^*,-1)x(0). \tag{1.17}$$

In problem (1.17) there are only rt^* variables $u(0)$, $u(1), \ldots, u(t^*-1)$, m linear equality constraints and $2rt^*$ linear constraints-inequalities. Now the expenditure matrix of problem (1.17) has already been filled up densely.

One can solve problem (1.17) by general linear programming methods. However, it should again be noted that these methods can substantially lose effectiveness at large t^* due to collinearity of vectors

$$HF(t^*,t)B(t),\ HF(t^*,\tau)B(\tau)$$

with close values t,τ.

In this chapter the form (1.17) of the optimal control problem (1.15) will be used for construction of an algorithm intended to account for all problem peculiarities and overcome its negative properties.

1.2. ε-MAXIMUM PRINCIPLE

The main result of qualitative theory of optimal control is the Pontryagin maximum principle discovered in 1956 for linear continuous systems. It provides necessary (and for linear problems also sufficient) conditions for optimality in extreme form. Below a constructive formulation and a generalization of the maximum principle on suboptimal controls are given.

1.2.1. Support control.

To simplify calculations we consider the case $r=1$. The following problem will be the basis of our investigations:

$$\sum_{t\in T} c(t)u(t) \longrightarrow max, \tag{2.1}$$

$$\sum_{t\in T} h(t)u(t)=\bar{g}, \quad f_*(t) \le u(t) \le f^*(t), \ t \in T$$

where

$$c(t) = c'F(t^*,t)b(t), \ h(t) = HF(t^*,t)b(t),$$
$$\bar{g} = g - F(t^*,-1)x(0). \tag{2.2}$$

As shown in Section 1.1.2 the last equation is equivalent to the terminal control problem.

The notation of problem (2.1) coincides with the canonical linear programming problem (Section 1.1.2) written in coordinate form

$$\sum_{j\in J} c_j z_j \longrightarrow max, \qquad \sum_{j\in J} a_j z_j = b, \ d_{*j} \le z_j \le d_j^*, \ j\in J. \tag{2.3}$$

Proceeding from this we shall modify our notions and constructions of the adaptive method for linear programming (see the Appendix) for optimal control problems. At the same time our attention will be given to their interpretation and convenient realization, reflecting the peculiarities of a new class of extremal problems.

The principle difference between problems (2.1) and (2.3) is as follows. Parameters c_j, a_j, d_{*j}, d^*_j, $j \in J$, b_i, $i \in I$, of problem (2.3) are considered to be arbitrary and the complete information about them is given by $|J|(|I| + + 3) + |I|$ numbers. Parameters $c(t)$, $h(t)$, $f_*(t)$, $f^*(t)$, $t \in T$, $\bar{g}$, of problem (2.1) according to (2.2) can be expressed using parameters c, $A(t)$, $b(t)$, g, $f_*(t)$, $f^*(t)$, $t \in T$, of problem (1.15). In the stationary case this information consists of $Ln + n^2 + mn + m + 2$ numbers. In the non-stationary case parameters $A(t)$, $b(t)$, $f_*(t)$, $f^*(t)$, $t \in T$, in their turn are often found to be functions of not a large number of independent parameters. In this context, parameters of problem (3.1) can be generated when necessary during the process of solving problem (1.15). The aim of realization of the adaptive method for the optimal control problem is to take into account peculiarities of problem (1.15).

Later on we use definitions from the Appendix.

Definition 2.1. The set of the moments $T_{sup} = \{ t_j \in T, j = \overline{1,m} \}$ is called a support of problem (1.15) if the $m \times m$ matrix

$$P = (h(t_j), t_j \in T_{sup})$$

is non-degenerate.

According to the results of Section 1.1 the support exists if and only if the dynamic system is controlled, i.e. the support is connected with one of the fundamental properties of the control systems.

The testing of whether the set T_{sup} is supportable is carried out in the following way. As seen from (2.2) , the

i-th component $h_i(t)$, $t \in T$, of $h(t)$, $t \in T$, is equal to

$$h_i(t) = \psi_i'^N(t)b(t), \; \psi_i'^N(t) = h'_{(i)}F(t^*,t)$$

where $h_{(i)}$ is the i-th row of H.

Using the equation for $F(t,\tau), \tau \in T$, we get

$$\psi_i'^N(t) = h'_{(i)}F(t^*,t) = h'_{(i)} F(t^*,t+1)A(t+1) =$$

$$= \psi_i'^N(t+1)A(t+1), \; i = \overline{1,m};$$

$$\psi_i^N(t^*-1)=h_{(i)}.$$

Therefore in testing if the chosen set $T_{sup} = \{ t_j \in T, j = \overline{1,m} \}$ is supportable we find m solutions $\psi_i^N(t)$, $t \in T$, $i = \overline{1,m}$ of the equation

$$\psi(t-1) = A'(t)\psi(t) \tag{2.4}$$

with

$$\psi_i \; (t^*-1) = h_{(i)}, \; i = \overline{1,m}. \tag{2.5}$$

Multiplying these solutions at t_j by the vectors $b(t_j)$ we get the columns

$$h(t_j) = (\psi_i^N \; b(t_j), \; i = \overline{1,m})$$

of $P = (h(t_j), \; j = \overline{1,m})$.

If $det \; P \neq 0$ then T_{sup} is a support.

By analogy with the support feasible point (see the Appendix) we introduce the following.

Definition 2.2. A pair $\{u, T_{sup}\}$ consisting of an admissible control and a support is called a support control.

We call a support control directly non-degenerate if its support components are non-critical, i.e.

$$f_*(t) < u(t) < f^*(t), \ t \in T_{sup}.$$

1.2.2. Vector of multipliers. Co-control. Conjugate system. Formula of increment of quality criterion.

In the Appendix the formula of increment

$$F(x) = - \sum_{j \in J_N} \Delta_j \Delta x_j, \quad \Delta_j = u' a_j - c_j, \quad u' = c'_{sup} Q,$$

$$c_{sup} = (c_j, j \in J_{sup}) \tag{2.6}$$

has been obtained.

Let $\upsilon = (\upsilon_i, i = 1,m)$ be the vector of multipliers for the optimal control problem. According to (2.6) it is equal to

$$\upsilon' = c'_{sup} Q \tag{2.7}$$

where $c_{sup} = (c(t), t \in T_{sup}), \ Q = P^{-1}$.

To calculate the components $c(t), t \in T_{sup}$, let us find the solution $\psi(t), t \in T$, of equation (2.4) with initial condition

$$\psi^c(t^* - 1) = c.$$

Then

$$c(t) = \psi^{c'}(t) b(t).$$

Let $\Delta(t)$, $t \in T_N$, $T_N = T\backslash T_{sup}$ be the estimates Δ_j, $j \in J_N$, for problem (1.16). From (2.6) we get

$$\Delta(t) = \upsilon' h(t) - c(t),\ t \in T_N.$$

Henceforth the function

$$\Delta(t) = \upsilon' h(t) - c(t),\ t \in T \tag{2.8}$$

is called a co-control (accompanying the support T_{sup}). It is the analogue of the accompanying co-point $\delta = \Delta$ from the Appendix.

We describe a new method of calculating co-controls. According to (2.2) we have:

$$\Delta(t) = \upsilon' HF(t^*, t)b(t) - c' F(t^*, t)b(t) =$$

$$= (H'\upsilon - c)F(t^*, t)b(t),\ t \in T.$$

We introduce the function

$$\psi'(t) = -(H'\upsilon - c)' F(t^*, t),\ t \in T.$$

Using equations (1.5),(1.7) we get

$$\psi'(t-1) = -(H'\upsilon - c)' F(t^*, t-1) = -(H'\upsilon - c)' F(t^*, t)A(t) =$$

$$= \psi'(t)A(t),\ t \in T,\ \psi(t^*-1) = c - H'\upsilon.$$

Thus the function $\psi(t), t \in T$, satisfies the equation

$$\psi(t-1) = A'(t)\psi(t) \tag{2.9}$$

with initial condition

$$\psi(t^*-1) = c - H'\upsilon. \tag{2.10}$$

Equation (2.9) is called a conjugate system. Its solution $\psi(t)$, $t \in T$, with condition (2.10) will be called the co-trajectory accompanying the support control $\{u, T_{sup}\}$. We get

$$\psi(t) = \psi^c(t) - \sum_{i=1}^{m} \upsilon_i \psi_i^N(t), \quad t \in T,$$

$$\Delta(t) = 0, \quad t \in T_{sup}.$$

Thus

$$\Delta(t) = -\psi'(t)b(t), \quad t \in T.$$

Then

$$\Delta J(u) = J(\bar{u}) - J(u) = -\sum_{t \in T} \Delta(t)\Delta u(t) =$$

$$= \sum_{t \in T} \psi'(t)b(t)\Delta u(t) \tag{2.11}$$

where $\Delta u(t) = \bar{u}(t) - u(t)$, $t \in T$, $\bar{u}(t)$, $t \in T$, is a control such that the corresponding trajectory $\bar{x}(t)$, $t \in T$, satisfies the terminal constraint $H\bar{x}(t^*)=g$.

Recalling the physical sense of the estimates Δ_j, $j \in J_N$, we can formulate the corresponding result for the co-control. At each $t \in T_N$ the number $\Delta(t)$ equals the change criterion speed of opposite sign under increasing the component $u(t)$. In addition the values of controls at all other non-support moments T_N are not changed and this control at support moments is chosen in such a way that the constraint $Hx(t^*) = g$ is fulfilled at the terminal moment.

1.2.3. Maximum principle.

The optimality criterion for problem (2.3) has the form

$$\Delta_j \geq 0 \text{ at } z_j = d_{*j}; \; \Delta_j \leq 0 \text{ at } z_j = d_j^*;$$
$$\Delta_j = 0 \text{ at } d_{*j} < z_j < d_j^*, \; j \in J_N.$$

In terms of co-control and co-trajectory we get

$$\Delta(t) = -\psi'(t)b(t) \geq 0 \text{ at } u(t) = f_*(t);$$

$$\Delta(t) = -\psi'(t)b(t) \leq 0 \text{ at } u(t) = f^*(t); \qquad (2.12)$$

$$\Delta(t) = -\psi'(t)b(t) = 0 \text{ at } f_*(t) < u(t) < f^*(t),\ t \in T_N.$$

We can write (2.12) in the extremal form

$$\psi'(t)b(t)u(t) = \max_{f_*(t)\leq u\leq f^*(t)} \psi'(t)b(t)u(t),\ t \in T_N. \qquad (2.13)$$

Let us introduce the function (Hamiltonian)

$$H(x,\psi,u,t)=\psi'(A(t)x+b(t)u).$$

In terms of the Hamiltonian, relation (2.13) can be written as follows:

$$H(x(t),\psi(t),u(t),t) = \max_{f_*(t)\leq u\leq f^*(t)} H(x(t),\psi(t),u(t),t),\quad t \in T_N. \qquad (2.14)$$

The optimality criterion of support controls can be formulated as follows.

Maximum principle. For optimality of the admissible control $u(t), t\in T$, it is sufficient to have a support T_{sup} such that along the support control $\{u,T_{sup}\}$ and the corresponding trajectories $x(t),\ t \in T \cup t^*$; $\psi(t),\ t \in T$, of systems (1.1),(2.9),(2.10) Hamiltonian attains maximum value (2.14).

If $\{u,T_{sup}\}$ is a non-degenerate support control then for optimality of the admissible control $u(t),\ t \in T$, it is necessary that along $\{u,T_{sup}\}$, $x(t),\ t \in T \cup t^*$; $\psi(t),\ t \in T$, relation (2.14) be fulfilled.

The difference between the given result and the discrete analogue of the Pontryagin maximum principle lies in the fact that it gives a concrete rule (2.10) for calculating the initial state $\psi(t^{*}-1)$ of the conjugate system (2.9). The Pontryagin maximum principle only declares the existence of nontrivial solution of the conjugate system with property (2.14). The addition given will make it possible to construct the effective algorithm for solution of the terminal problem of optimal control.

The testing of the maximum principle of is performed by the following "dynamic" method.

Using the support control $\{u, T_{sup}\}$ the vector of multipliers (2.7) is calculated. From the right to the left the conjugate system (2.9),(2.10) is solved. At the same time the scalar product $\psi'(t)b(t)$ is set up and the relations (2.14) are tested. In the process of testing the problem parameters (2.1) are not used. Necessary values are generated from parameters of problem (1.15) by a special (dynamic) method.

1.2.4. ε-maximum principle

Using the criterion of suboptimality of a support feasible point (see the Appendix) we get

$$\beta(x, J_{sup}) = \sum_{j \in J_N} \Delta_j (x_j - \kappa_j) \leq \varepsilon \tag{2.15}$$

where $\kappa_j, j \in J_N$, are components of the accompanying pseudo-feasible point

$$\kappa_j = d_{*j} \text{ at } j \in J_N^+, \; \kappa_j = d_j^* \text{ at } j \in J_N^- . \tag{2.16}$$

In terms of the optimal control problem (1.15) inequality (2.15) and relation (2.16) takes the form

$$\sum_{t\in T_N}\Delta(t)(u(t)-\omega(t)) = \sum_{t\in T_N}\psi'(t)b(t)(\omega(t)-u(t)) \le \varepsilon. \qquad (2.17)$$

Here $\omega(t), t\in T$, is a pseudo-control. It is equal to

$$\omega(t) = f_*(t),\ t \in T_N^+ = \{ t \in T_N:\ \Delta(t) \ge 0 \},$$

$$\omega(t) = f^*(t),\ t \in T_N^- = T_N \backslash T_N^+ = \{ t \in T_N:\ \Delta(t) \le 0 \}$$

and satisfies condition

$$\psi'(t)b(t)\omega(t) = \max_{f_*(t)\le u\le f^*(t)} \psi'(t)b(t)u,\ t \in T_N, \qquad (2.18)$$

$$\omega(T_{sup}) = Q(g-H\kappa_N(t^*))$$

where $\kappa_N(t^*)$ is the terminal state of system (1.1), corresponding to the control

$$u(t) = \omega(t),\ t \in T_N,\ u(t) = 0,\ t \in T_{sup}.$$

Using (2.18) and $H(x,\psi,u,t) = \psi'(A(t)x+b(t)u)$, we can write inequality (2.17) as

$$H(x(t),\psi(t),u(t),t) =$$

$$= \max_{f_*(t)\le u\le f^*(t)} H(x(t),\psi(t),u(t),t)-\varepsilon(t), t\in T_N, \qquad (2.19)$$

$$\beta(u,T_{sup}) = \sum_{j\in T_N} \varepsilon(t) \le \varepsilon . \qquad (2.20)$$

Now the criterion of suboptimality can be formulated as follows.

ε-maximum principle. At any $\varepsilon \ge 0$, for ε-optimality of the admissible control $u(t),\ t \in T$, it is necessary and sufficient to have such support T_{sup} that along $\{u,T_{sup}\}$ and $x(t),\ t \in T \cup \cup t^*$; $\psi(t),\ t \in T$, the Hamiltonian of the problem attains the ε - maximum value (2.19).

The testing of the ε-maximum principle is performed by the same dynamic method as the testing of the maximum principle.

As in the case with the feasible support point $\{x, J_{sup}\}$ for linear programming problems (see the Appendix) we introduce measures of non-optimality of the support control components $\{u, T_{sup}\}$. The number

$$\beta(u) = J(u^0) - J(u) = c'x^0(t^*) - c'x(t^*)$$

is said to be the measure of non-optimality of the admissible control $u(t)$, $t \in T$.

By analogy with

$$\beta(J_{sup}) = c'\kappa - c'x^0$$

the number

$$\beta(T_{sup}) = J(\omega) - J(u^0) = c'\kappa(t^*) - c'x^0(t^*)$$

is said to be the measure of non-optimality of the support T_{sup} where $\kappa(t)$, $t \in T \cup t^*$, is the (pseudo-trajectory) trajectory of system (1.1) corresponding to the pseudo-control $\omega(t)$, $t \in T$.

The estimate of suboptimality $\beta(u, T_{sup})$ (2.20) has a decomposition

$$\beta(u, T_{sup}) = \beta(u) + \beta(T_{sup}) = c'\kappa(t^*) - c'x(t^*).$$

Example 2.1. We consider the discrete analogue of the optimal control problem (mechanical motion):

$$J(u) = x_1(5) \longrightarrow \max, \ x_1(t+1) = x_1(t) + x_2(t),$$

$$x_2(t+1) = x_2(t) + u(t), \ x_1(0) = x_2(0) = 0, \qquad (2.21)$$

$$x_2(5) = 0, \quad u(t) \le 1, \ t \in T = \{0,1,2,3,4\}.$$

Problem (2.21) is the specific case of problem (1.15)

$$c = \begin{bmatrix} 1 \\ 0 \end{bmatrix}, \quad A(t) = \begin{bmatrix} 1 & 1 \\ 0 & 1 \end{bmatrix}, \quad b(t) = \begin{bmatrix} 0 \\ 1 \end{bmatrix}, \quad H = [0 \ 1], \ g = 0,$$

$$f^{*}(t) = 1, \ f_{*}(t) = -1, \ t \in T.$$

Consider the admissible control $u(t) = 0, \ t \in T$, corresponding to the system trajectory $x(t) = 0, \ t = 1, \ldots, 5$. We get $J(u) = 0$.

Assign the support $T_{sup} = \{ t^{*}-1 \} = \{ 4 \}$ to the control $u(t) = 0, \ t \in T$. The support control $\{ u, T_{sup} \}$ is non-degenerate, i.e. $-1 \le u(4) = 0 < 1$.

Since $T_{sup} = \{t^{*}-1\}$, to construct the support matrix $P = (h(t), \ t \in T_{sup})$ it is sufficient to find the state of conjugate system

$$\psi^{N}(t - 1) = \begin{bmatrix} 1 & 0 \\ 1 & 1 \end{bmatrix} \psi^{N}(t), \quad \psi^{N}(t^{*}-1) = \psi(4) = \begin{bmatrix} 0 \\ 1 \end{bmatrix}$$

at one moment $t = t^{*}-1 = 4$:

$$P = (h(t^{*}-1)) = (h(4)) = (\psi^{N'}(4)b(4)) = [0 \ 1]\begin{bmatrix} 0 \\ 1 \end{bmatrix} = [1].$$

The support component c_{sup} is also simply calculated:

$$c_{sup} = (c(t^{*}-1)) = (\psi^{c'}(t^{*}-1)b(t^{*}-1)) =$$

$$= (\psi^{c'}(4)b(4)) = [\ 0 \ 1 \ ]\begin{bmatrix} 0 \\ 1 \end{bmatrix} = [1].$$

The vector of multipliers equals

$$\nu' = c_{sup}' Q = 0 \times 1 = 0.$$

We find the co-trajectory

$$\psi(t-1) = \begin{bmatrix} 1 & 0 \\ 1 & 1 \end{bmatrix} \psi(t), \quad \psi(t^*-1) = \psi(4) = c - H'\upsilon = \begin{bmatrix} 1 \\ 0 \end{bmatrix}$$

and

$$\psi(3) = \begin{bmatrix} 1 \\ 1 \end{bmatrix}, \psi(2) = \begin{bmatrix} 1 \\ 2 \end{bmatrix}, \psi(1) = \begin{bmatrix} 1 \\ 3 \end{bmatrix}, \psi(0) = \begin{bmatrix} 1 \\ 4 \end{bmatrix}.$$

Calculate the co-control

$$\Delta(t) = -\psi'(t)b(t), \quad t \in T;$$

$$\Delta(0) = -4, \ \Delta(1) = -3, \ \Delta(2) = -2, \ \Delta(3) = -1, \ \Delta(4) = 0.$$

Test the maximum principle for the support control $\{u, T_{sup}\}$:

$$\Delta(0) = -4 \quad \text{if} \quad -1 < u(0) = 0 < 1;$$

$$\Delta(1) = -3 \quad \text{if} \quad -1 < u(1) = 0 < 1;$$

$$\Delta(2) = -2 \quad \text{if} \quad -1 < u(2) = 0 < 1;$$

$$\Delta(3) = -1 \quad \text{if} \quad -1 < u(3) = 0 < 1.$$

The last equation means that the maximum principle is not fulfilled.

To calculate the estimate of suboptimality of the support control $\{u, T_{sup}\}$ let us find non-support components of the pseudo-control $\omega(t)$, $t \in T_N$. Since $\Delta(t) < 0$, $t \in T_N$, then $\omega(t) = f^*(t) = 1$, $t \in T_N$, i.e.

$$\omega(0) = \omega(1) = \omega(2) = \omega(3) = 1.$$

According to (2.17) the estimate of suboptimality is equal to

$$\beta(u, T_{sup}) = \sum_{t \in T_N} \Delta(t)(u(t) - \omega(t)) = 10.$$

We find that the maximum value $J(u)$ does not exceed $J(u) + \beta(u, T_{sup}) = 10$.

1.3. ALGORITHM FOR CONSTRUCTING OPTIMAL PROGRAM CONTROLS.

After "translation" into terms of optimal control basic concepts and facts which are related to adaptive linear programming method it remains to apply the algorithm itself to a new class of extremal problems.

Before performing iterations we assume that the initial support control $\{ u,T_{sup} \}$ and the accuracy of approximation $\varepsilon \geq 0$ to the control with respect to the criterion are given.

1.3.1. Substitution of an admissible control.

Let $\{ u, T_{sup} \}$ be an initial support control. We calculate the vector of multipliers $\upsilon' = c'_{sup}Q$ and construct the co-trajectory $\psi(t), t\in T$:

$$\psi(t-1) = A'(t)\psi(t) \ , \ \psi(t^{*}-1) = c - H'\upsilon,$$

the co-control

$$\Delta(t) = -\psi'(t)b(t) \ , \ t \in T$$

and the non-support components of the pseudo-control

$$\omega(t) = f_{*}(t), \ t \in T_N^{+} = \{ t \in T_N : \Delta(t) \geq 0\},$$

$$\omega(t) = f^{*}(t), \ t \in T_N^{-} = T_N \setminus T_N^{+} = \{ t \in T_N : \Delta(t) \leq 0 \}.$$

We obtain the estimate of suboptimality

$$\beta(u,T_{sup}) = \sum_{t\in T_N} \Delta(t)(u(t)-\omega(t)).$$

If $\beta(u,T_{sup}) \leq \varepsilon$, then u is an ε-optimal control and the solution process is complete. Otherwise the solution process is continued.

Let us calculate the support components of the pseudo-control (Section 1.2):

$$\omega_{sup} = (\ \omega(t),\ t \in T_{sup}\ ) = Q(g-H\kappa_N(t^*))$$

and find the value μ which is equal to their maximum outcome beyond the constraints (1.13)

$$\mu = max\ \rho(\omega(t),\ [\ f_*(t),\ f^*(t)]\ )\ ,\ t \in T_{sup}.$$

If $\mu=0$, i.e.

$$f_*(t) \le \omega(t) \le f^*(t),\ t\in T_{sup} \tag{3.1}$$

then $\omega(t),\ t \in T,$ is an optimal control and the solution process is complete.

At $\mu \le \mu_0$ where $\mu_0 \ge 0$ is a parameter of the method, we substitute the support T_{sup} (see Section 1.3.2).

We consider the case when $\mu > \mu_0$. We construct direction $\Delta u(t),\ t \in T,$ of control variation

$$\Delta u(t)= u(t) - \omega(t),\ t \in T$$

and calculate the maximum admissible step ϑ^0 along this direction,

$$\vartheta^0 = \vartheta(t_0) = \ min\ \ \vartheta(t),\ t \in T_{sup},$$

$$\vartheta(t) = \begin{cases} (f^*(t)-u(t))/\Delta u(t) \text{ at } \Delta u(t) > 0, \\ (f_*(t)-u(t))/\Delta u(t) \text{ at } \Delta u(t) < 0, \\ \quad \infty \quad \text{ at } \Delta u(t)=0. \end{cases}$$

Construct a new admissible control:

$$\bar{u}(t) = u(t) + \vartheta^0 \Delta u(t),\ t \in T.$$

The estimate of suboptimality of the support control $\{\bar{u}, T_{sup}\}$ is equal to

$$\beta(\bar{u}, T_{sup}) = (1-\vartheta^0)\beta(u, T_{sup}).$$

At $\beta(\bar{u}, T_{sup}) \le \varepsilon$, we stop the process of solution at the ε-optimal control $\bar{u}(t)$, $t \in T$. If $\beta(\bar{u}, T_{sup}) > \varepsilon$, we start the substitution of the support.

1.3.2. Substitution of a support.

We make the substitution of a support according to the "long step" rule (see the Appendix). It takes into account the specifics of the optimal control problem more completely than the "short step" rule.

Let us rewrite operations concerning the substitution of a support in terms of the optimal control problem.

Construct the direction $\Delta\delta(t)$, $t \in T$, of co-control variation

$$\Delta\delta(t) = \Delta\psi'(t)b(t), \quad t \in T$$

where $\Delta\psi(t)$, $t \in T$, is the trajectory of the conjugate system

$$\psi(t-1) = A'(t)\psi(t)$$

with initial condition

$$\Delta\psi(t^*-1) = H'\Delta\upsilon,$$

$\Delta\upsilon \in R^m$ is the variation direction for the vector of multipliers satisfying

$$\Delta\delta(t) = 0, \ t \in T_{sup} \backslash t_0; \ \Delta\delta(t_0) = -\mathrm{sign}\, \Delta u(t_0).$$

It is equal to

$$\Delta v = -\ q(t_o) sign\ \Delta u(t_o)$$

where $q(t_o)$ is the line of the matrix

$$Q = Q(T_{sup}, I) = P(I, T_{sup})^{-1}$$

corresponding to t_o.

Calculate

$$\gamma_o = -|\omega(t_o) - \bar{u}(t_o)| \ .$$

For each $t \in T_N$ we have

$$\sigma(t) = \begin{cases} -\Delta(t)/\Delta\delta(t) & \text{at } \Delta(t)/\Delta\delta(t) < 0 \text{ or } \Delta(t)=0 \text{ and} \\ \omega(t)=f^*(t),\ \Delta\delta(t) > 0 \text{ or } \Delta(t)=0 \text{ and} \\ \omega(t)=f_*(t),\ \Delta\delta(t) < 0 \text{ or } \Delta(t)=0; & \infty \text{ in other cases;} \end{cases}$$

$$\Delta\gamma(t) = |\Delta\delta(t)|(f^*(t)-f_*(t)),\ t \in T_N; \quad \sigma(t) \neq \infty.$$

Put the moments $t \in T_N$, $\sigma(t) \neq \infty$, in the increasing order

$$\sigma(t) : \ t_1, t_2, \ldots, t_p;\ \sigma(t_i) \le \sigma(t_{i+1}),\ i = \overline{1,p-1}$$

and find moment t_q such that

$$\Delta\gamma(t_q) = \gamma_o + \sum_{i=1}^{q} \Delta\gamma(t_i) \ge 0,\ \gamma(t_{q-1}) < 0. \qquad (3.2)$$

Suppose

$$T_{sup} = (\ T_{sup} \setminus t_o\ ) \cup t_*,\ t_* = t_q.$$

We pass to the new iteration with the support control $\{\ \bar{u}, \bar{T}_{sup}\}$. Using the results of the Appendix we calculate

$$\bar{Q} = \bar{P}^{-1} = (h(t),\ t\in\bar{T}_{sup})^{-1}$$

and the auxiliary vector

$$r = r(T_{sup}) = Qh(t^*) = Qh\bar{x}(t^*)$$

where $\bar{x}(t),\ t \in T \cup t^*$, is the trajectory of system

$$\bar{x}(t+1)=A(t)\bar{x}(t)$$

with initial condition

$$x(t_*)=b(t_*).$$

The procedure is repeated.

Example 3.1. Let us fulfil one iteration of the adaptive method for the control support

$$\{u,T_{sup}\},u(t) = 0,\ t \in T,\ T_{sup} = \{\ 4\ \}.$$

It has been calculated (see Example 2.1) that

$$Q = 1,\ \upsilon = 0,\quad \Delta(t) = (\ -4,\ -3,\ -2,\ -1,\ 0),$$

$$\omega(0) = \omega(1) = \omega(2) = \omega(3) = 1,$$

$$\beta(u,T_{sup})=10.$$

We begin the iteration through the substitution of a feasible point. For this purpose we calculate support components of the pseudo-control

$$\omega_{sup} = \omega(4) = Q(g-Hx_N(t^*))$$

where $x_N(t)\ t \in T$, is a trajectory of system

$$x(t+1) = \begin{bmatrix}1 & 1\\ 0 & 1\end{bmatrix}x(t) + \begin{bmatrix}0\\ 1\end{bmatrix}u(t),\ x_N(0)=0\ ,$$

and

$$u = (\ \omega(t),\ t \in T_N;\ \omega(t) = 0,\ t \in T_{sup}\ ) =$$

$$= (\ 1,\ 1,\ 1,\ 1,\ 0).$$

We get

$$x_N(1) = \begin{bmatrix} 0 \\ 1 \end{bmatrix}, x_N(2) = \begin{bmatrix} 1 \\ 2 \end{bmatrix}, x_N(3) = \begin{bmatrix} 3 \\ 3 \end{bmatrix}, x_N(4) = \begin{bmatrix} 6 \\ 4 \end{bmatrix}, x_N(5) = \begin{bmatrix} 10 \\ 4 \end{bmatrix}.$$

Hence $\omega(4) = 1(0 - [0\ 1]\begin{bmatrix} 10 \\ 4 \end{bmatrix}) = -4.$

Since $\omega(4) < f_*(4) = -1$, we continue the solution process.

Let us construct

$$\Delta u(t) = \omega(t) - u(t),\ t \in T:$$

$$(\ \Delta u(t),\ t \in T) = (\ 1,\ 1,\ 1,\ 1,\ -4\ )$$

and calculate

$$\vartheta^0 = \min\ \{\ \vartheta(t),\ t \in T_{sup}\} = \vartheta(4) =$$

$$= -\frac{f_*(4) - u(4)}{\Delta u(4)} = 1/4\ ; \qquad t_0 = 4.$$

The new control has the form $\bar{u} = u + \vartheta^0 \Delta u = (\ 1/4,\ 1/4,\ 1/4,\ 1/4,\ -1)$. The suboptimality estimate of the support control $\{\ \bar{u}, T_{sup}\ \}$ is

$$\beta(\bar{u}, T_{sup}) = (1 - \vartheta^0)\beta(u, T_{sup}) = 3/4 \times 10 = 7.5.$$

We pass to the substitution of the support. Having calculated

$$\gamma_0 = -|\bar{u}(t_0) - \omega(t_0)| = -3;$$

$$\Delta\upsilon = \Delta\upsilon_1 = -q(t_0)sign\ \Delta u(t_0) = 1$$

and integrated the conjugate system with

$$\Delta\psi(t^*-1) = \Delta\psi(4) = H'\Delta\upsilon = \begin{bmatrix} 0 \\ 1 \end{bmatrix}$$

we get

$$\Delta\psi(3) = \begin{bmatrix} 0 \\ 1 \end{bmatrix},\ \Delta\psi(2) = \begin{bmatrix} 0 \\ 1 \end{bmatrix},\ \Delta\psi(1) = \begin{bmatrix} 0 \\ 1 \end{bmatrix},\ \Delta\psi(0) = \begin{bmatrix} 0 \\ 1 \end{bmatrix},$$

and

$$\Delta\delta(t) = \Delta\psi'(t)b(t)\ ,\ t \in T:\ \Delta\delta(0) = 1,\ \Delta\delta(1) = 1,$$

$$\Delta\delta(2) = 1,\ \Delta\delta(3)=1,\ \Delta\delta(4)=1.$$

For $t \in T_N = \{\ 0,\ 1,\ 2,\ 3\ \}$ we calculate

$$\sigma(0) = -\ \Delta(0)/\Delta\delta(0) = 4, \sigma(1) = 3,\ \sigma(2) = 2,\ \sigma(3) = 1,$$

$$\Delta\gamma(0) = 2,\ \Delta\gamma(1) = 2,\ \Delta\gamma(2) = 2,\ \Delta\gamma(3)=2.$$

As

$$\sigma(3) < \sigma(2) < \sigma(1) < \sigma(0)$$

and

$$\gamma(3)=\gamma_0+\Delta\gamma(3)=-1; \gamma(2)=\gamma(3)+\Delta\gamma(2)=1,$$

then $t_* = 2$.

We construct the new support

$$\bar{T}_{sup} = (\ T_{sup} \setminus t_0\ ) \cup t_* = \{\ 2\ \},\ \bar{Q} = (1).$$

The suboptimality estimate of the new support control $\{\bar{u}, T_{sup}\}$ equals

$$\beta(\bar{u},\bar{T}_{sup}) = \beta(\bar{u},T_{sup}) + \gamma_0\sigma(t_1) + \gamma(t_1)(\sigma(t_2)-\sigma(t_1))+\ldots$$

$$+\ \gamma(t_{q-1})(\sigma(t_q) - \sigma(t_{q-1})) = 7.5 + (-\ 3)\times 1 + (-1)\times(2-1) = 3.5.$$

The quality criterion on the control $\bar{u}(t),\ t \in T,$

takes the value $J(\bar{u}) = c'x(t^*) = 2.5$ This value differs from the optimal value by not more than 3.5.

It can be shown that $\bar{T}_{sup}$ is the optimal support ,

$$J(u^0) = 6 \ , \ u^0 = (\ 1,\ 1,\ 0,\ -1,\ -1).$$

At the phase plane of variables (x_1, x_2) the optimal trajectory takes the form presented in Fig.3.1. The trajectory $\bar{x}(t)$, $t \in T$, which is obtained after change of the initial control is also presented in Fig. 3.1.

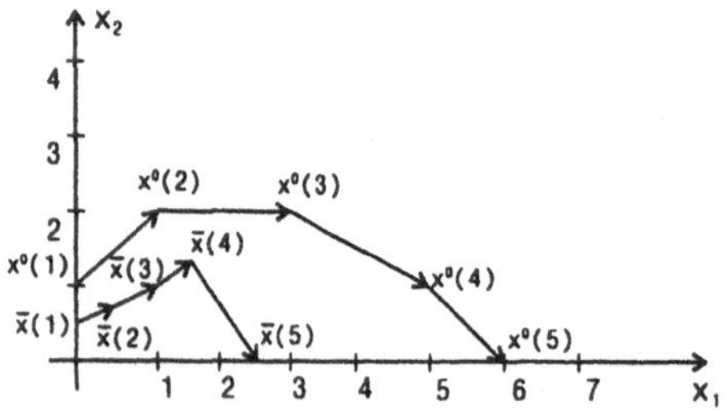

Fig.3.1

CHAPTER 2

DYNAMIC UNCERTAIN SYSTEMS

2.1. ADJOINT PROBLEMS OF CONTROL, OBSERVATION AND IDENTIFICATION.

Problems of control and observation of dynamic systems, dual in Kalman's sense, are well known in the qualitative theory of optimal processes. The duality principle is used in many papers and is a general one in the procedure of observation [24,30].

We suggest relating the problems of control, observation and identification with the help of extremal problems which can be reduced to a pair comprising either an optimal control problem and an observation problem or an optimal control problem and an identification problem. This connection can be used to construct some elements of feedback in dynamic systems.

2.1.1. The observation problem adjoint to the problem of optimal control of initial states.

Let the set

$$X_0 = \{ x \in \mathbb{R}^n : Dx = b,\ d_* \le x \le d^* \},$$

$$\{ d_*,\ d^* \in \mathbb{R}^n;\ b \in \mathbb{R}^m;\ D \in \mathbb{R}^{n\times m} \},$$

a piecewise continuous function $y(t)$, $t \in T = [0, t^*]$, an $n \times n$-matrix D, vectors $c, h \in \mathbb{R}^n$, and scalars g_*, g^* be given.

We consider the problem of optimal choice of initial states of dynamic system in which the motion $y(t)$, $t \in T$, is realized, i.e.

$$c'z \longrightarrow max \ , \tag{1.1}$$

$$\dot{x} = Ax, \ x(0) = z \ , \tag{1.2}$$

$$z \in X_0 \ , \tag{1.3}$$

$$g_* \le y(t) - h'x(t) \le g^*, \ t \in T. \tag{1.4}$$

The adjoint observation problem for (1.1) - (1.4) is formulated in the following way.

We suppose that the a priori distribution of initial states X_0 of the dynamic system (1.2) is known. Let us use the device

$$y = h'x + \xi \tag{1.5}$$

that measures output signals $h'x(t)$, $t \in T$, with some error $\xi(t)$, $t \in T$. The error realizations $\xi(t)$, $t \in T$, are assumed to be piecewise continuous functions satisfying conditions

$$g_* \le \xi(t) \le g^*, \ t \in T. \tag{1.6}$$

The set X^* of initial states $z \in X_0$ .which can generate the observed signal $y(t)$, $t \in T$, together with some errors $\xi(t)$, $t \in T$, is called an a posteriori distribution of the initial states of system (1.2).

The simplest number characteristic of the set X^* is the extension $max\ c'z$, $z \in X^*$, in the direction c (c is given).

The observation problem

$$c'z \longrightarrow max, \ z \in X^*,$$

will be called an adjoint problem with respect to the optimal control problem (1.1) - (1.4). Both problems are reduced to the semi-infinite extremal problem:

$$c'z \longrightarrow max, \quad b_*(t) \le a'(t)z \le b^*(t), \; t \in T;$$

$$Dz = b, \; d_* \le z \le d^* . \tag{1.7}$$

with a finite number of variables and infinite number of constraints. Indeed for problem (1.1) - (1.4) we get

$$x(t) = F(t, O)z \;, \; z \in X_o \;,$$

$$g_* \le y(t) - h'F(t,o)z \le g^*$$

where $F(t, t_o)$ is the fundamental matrix of solutions of the system

$$\dot{x} = Ax \;,$$

i.e. the problem

$$c'z \longrightarrow max, \quad b_*(t) \le a'(t)z \le b^*(t), \; t \in T;$$

$$x(t) = F(t,O)z, \; z \in X_o \tag{1.8}$$

arises. We shall get the same problem if the above formulated adjoint observation problem is considered. Really, the observation problem has the form

$$c'z \longrightarrow max, \; z \in \hat{X}_o \;,$$

$$y = h'x + \xi \;, \tag{1.9}$$

$$x(t) = F(t, O)z \;,$$

Using (1.6) we get

$$g_* \le y(t) - h'F(t, o)z \le g^*,$$

($y(t)$ is some particular function). So we have come up to (1.8).

2.1.2. The problem of identification of perturbations adjoint to the optimal control problem.

Let in addition a piecewise continuous function $c(t)$, $t \in T$; n-vectors b, x_0 and a scalar y^* be given.

We shall consider the optimal control problem

$$\int_0^{t^*} c(t)u(t)dt \longrightarrow max \ , \quad \dot{x} = Ax + bu \ , \quad x(0) = x_0 \ , \tag{1.10}$$

$$g_* \le y - h'x(t^*) \le g^* , \quad |u(t)| \le 1, \ t \in T,$$

in the class of piecewise continuous functions $u(t)$, $t \in T$.

Construct an adjoint observation problem with respect to (1.10). Let a priori distribution Ω_0 of piecewise continuous perturbations $w(t)$, $t \in T$, be known, and the perturbations $w(t)$, $t \in T$, act as an input of the dynamic system:

$$\dot{x} = Ax + bw(t), \ x(0) = x_0 \ , \tag{1.11}$$

$$w(t) \in \Omega_0 = \{ \ w(t), \ t \in T: \ |w(t)| \le 1 \ \}.$$

The measuring device (1.5) indicates the signal $h'x(t^*)$ at the moment t^* with an error ξ that can have values satisfying the conditions

$$g_* \le \xi \le g^* . \tag{1.12}$$

Denote the a posteriori distribution of perturbations by Ω^*. It consists of those and only those $w(t)$, $t \in T$, that can generate the observed signal y^* with error ξ. The simplest number characteristic of the set Ω^* is the maximum meaning of the moment

$$\int_0^{t^*} c(t)w(t)dt$$

The calculation of this characteristic is reduced to the following extremal problem:

$$\int_0^{t^*} c(t)w(t)dt \longrightarrow \max_W, \quad w(\cdot) \in \Omega^*. \qquad (1.13)$$

Problem (1.13) will be called a perturbation identification problem adjoint with respect to the optimal control problem (1.10).

Problems (1.8) - (1.11) are particular cases of the semi-infinite extremal problem

$$\int_0^{t^*} c(t)u(t)dt \longrightarrow \max, \quad b_* \le \int_0^{t^*} a(t)u(t)dt \le b^*, \qquad (1.14)$$

$$|u(t)| \le 1, \; t \in T,$$

with infinite number of variables and finite number of general constraints.

2.1.3. The observation problem adjoint to the problem of optimal control of terminal states with phase constraints.

Let us consider the terminal problem of constructing optimal control:

$$c'x(t^*) \longrightarrow \max, \quad \dot{x} = Ax + bu, \quad x(0) = x_0 ,$$

$$g_* \le y(t)-h'x(t) \le g^*, \quad |u(t)| \le 1, \quad t \in T$$

where $y(t)$, $t \in T$, is a given function.

This problem is connected to the following adjoint observation problem:

$$c'x \longrightarrow \max, \quad x \in X^*(t^*)$$

intented to calculating the extension of a posteriori distribution $X^*(t^*)$ of terminal states $x(t^*)$ of system (1.11) under assumption that the system is affected by piecewise linear perturbations $w(t)$, $t \in T$,

$$|w(t)| \le 1, \quad t \in T .$$

We suppose that the signal $y(t)$, $t \in T$, is obtained by the measuring device (1.5) with errors satisfying (1.6).

The mentioned problems of control and observation are particular cases of the infinite extremal problem

$$\int_0^{t^*} c(t)u(t)dt \longrightarrow max, \quad b_*(t) \le \int_0^{t^*} a(t,\tau)u(\tau)d\tau \le b^*(t) , \tag{1.15}$$

$$|u(t)| \le 1, \quad t \in T.$$

Finite algorithms for solving problems (1.7), (1.14), (1.15) and their application to solving various problems of optimal control are presented in [19]. It is important to emphasize that according to results of Sections 2.1.1.– 2.1.3. these algorithms can also be used for solving observation problems.

2.1.4. The identification problem adjoint to the problem of optimal control of dynamic system parameters.

Let $n \times n$ matrices $A_0, A_1, \ldots, A_q$; n-vectors $b_0, b_1, \ldots, b_q$; piecewise continuous functions $u_0(t), u_1(t), \ldots, u_q(t)$, $t \in T$, and a set

$$W = \{ w \in \mathbb{R}^q : Gw = f, w_* \le w \le w^* \}, \; (f \in \mathbb{R}^l) , \tag{1.16}$$

of values of parameter $w = (w_1, \ldots, w_q)$, on which the matrix A, the vector b and the control $u(t)$, $t \in T$, of the dynamic system

$$\dot{x} = Ax + bu, \quad x(0) = x_0,$$

$$A = A_0 + \sum_{i=1}^{q} w_i A_i \quad b = b_0 + \sum_{i=1}^{q} w_i b_i \ , \tag{1.17}$$

$$u(t) = u_0(t) + \sum_{i=1}^{q} w_i u_i(t), \quad t \in T,$$

depend, be given.

We consider the problem of optimal control of the dynamic system parameters

$$c'x \longrightarrow max,$$

$$\dot{x} = (A_0 + \sum_{i=1}^{q} w_i A_i)x + (b_0 + \sum_{i=1}^{q} w_i b_i)(u_0(t) + \tag{1.18}$$

$$+ \sum_{i=1}^{q} w_i u_i(t)),$$

$$x(0) = x_0, \quad t \in T; \quad g_* \le y - h'x(t^*) \le g^*, \quad w \in W.$$

Let us formulate the problem of identification adjoint with respect to problem (1.18).

Let the dynamic system (1.17) with an unknown value of vector w be functioning at the segment T. A priori distribution W (1.16) of the parameters is known. To make this distribution more correct at t^*, the signal $y^*=y(t^*)$ is written by the measuring device (1.5). A posteriori distribution W^* consist of only those parameters $w \in W$ which can generate an output signal y^* together with errors $\xi=\xi(t^*)$.

Calculation of the extension of the set W^* along the direction c:

$$c'w \longrightarrow max, \quad w \in W^*, \tag{1.19}$$

is called a problem of identification of the dynamic system.

It is adjoint with respect to the optimal control problem (1.19).

Problems (1.18) - (1.19) represent particular cases of the nonlinear programming problem

$$c'x \longrightarrow max, \quad f(x) = 0, \quad d_{*} \leq x \leq d^{*}. \tag{1.20}$$

Special methods for solving problem (1.21) based on network models of nonlinear functions are elaborated in [19].

One method of improving the structures of the a posteriori distributions X^{*}, $X^{*}(t^{*})$, Ω^{*}, W^{*} more correct is to introduce quantization of these sets in various ways.

We choise above the four special problems of control, observation and identification, only for simplicity and definiteness. Their various generalizations are evident. It is important to notice that according to the described approach all these problems are primal. It is known that the primal form for control problems and the dual form for observation problems are widely used. From the point of view of the constructive theory of extremal problems, primal forms are more natural. Dual forms play an important role but only as auxiliary problems.

2.2. A FINITE ALGORITHM FOR CONSTRUCTING PROGRAM SOLUTIONS OF AN INCOMPLETELY DETERMINED LINEAR OPTIMAL CONTROL PROBLEM.

The classical theory of optimal control was first constructed for determined problems [6,13,38]. Later on two approaches to optimal control problems under conditions of uncertainty were outlined. In the first approach random elements were considered along with determined elements. The theory of stochastic processes developed intensively in the postwar years served as its basis [2,14,31,39]. Another approach was aimed at obtaining guaranteed results [27-30,40]. It described uncertainty only by sets of feasible values of undetermined elements of the problem without introduction of probability measurement. The development of the second approach is connected to the modern theory of extremal problems.

Methods of solving various extremal problems with constraints were worked out by the authors between 1975 and 1991 in [15-22]. On the basis of these results the method of dynamic system optimization under conditions of uncertainty is considered below.

2.2.1. Problem statement.

We consider the dynamic system

$$x = Ax + bu \quad (x \in \mathrm{R}^n, \ u \in \mathrm{R}) \tag{2.1}$$

in the class of piecewise continuous controls $u(t)$, $t \in T = [0,t^*]$ with constraints

$$|u(t)| \le 1, \ t \in T, \tag{2.2}$$

with incompletely determined initial state

$$x(0) \in \check{X}_0 = \{ x \in R^n : Gx = f,\ d_* \le x \le d^* \}. \tag{2.3}$$

Here G is a given $q \times n$ matrix, $q \le n$, d_*, d^* are given n-vectors.

The set $\check{X}_0$ will be called the a priori distribution of initial states. This set and every control $u(\cdot) = (u(t),\ t \in T)$ are put into correspondence with

$$\check{X}(t) = X(t|u(\ )) = \{ x(t|x_0, u(\cdot)),\ x_0 \in \check{X}_0 \},\ t \in T.$$

It is the change of the a priori state distribution that will further be called an a priori movement of system (2.1).

Let the terminal set

$$X^* = \{ x \in R^n : h_i' x \ge g_i,\ i = \overline{1,m} \} \tag{2.4}$$

be defined in state space.

The a priori movement $\check{X}(t|\check{u}(\cdot))$, $t \in T$, and the control $\check{u}(\cdot)$ (2.2) generating it will be called a priori admissible if they satisfy the terminal inclusion

$$\check{X}(t^*|\check{u}(\cdot)) \subset X^*. \tag{2.5}$$

We define the quality criterion of control

$$J(\check{u}) = \min h_0' x(t^*|x_0, \check{u}(\cdot)),\ x_0 \in \check{X}_0, \tag{2.6}$$

on a priori admissible movements.

A priori admissible control $\check{u}^0(t)$, $t \in T$, will be called a priori optimal if

$$J(\check{u}^0) = \max_{\check{u}(\cdot)} J(\check{u}). \tag{2.7}$$

It is obvious that the uncertainty (2.3) can only decrease the control efficiency (2.7). To increase the efficiency it is worth while decreasing the degree of uncertainty. A way to achieve this purpose is observation and analysis of control processes.

Observation may be accomplished by using complete, incomplete, accurate (without errors) or incomplete (with errors) measurements. Inertia free complete measurements are accomplished by device $y=Px$, where P is the non-degenerate $n\times n$ matrix[1)*]. For incomplete accurate measurements the measuring device

$$y = Kx \tag{2.8}$$

is used, where K is an $l\times n$ matrix, $l \leq n$, $rank\ K = l$.

An inertia free measuring device accomplishing incomplete measurements with errors is described by

$$y = Kx + \xi, \tag{2.9}$$

We assume that the measurement errors $\xi(t)$, $t \in T$, can turn out any piecewise continuous l-vector functions satisfying unequalities:

$$\xi_* \leq \xi(t) \leq \xi^*, \ t\in T. \tag{2.10}$$

Assume that the chosen control $u(t)$, $t\in T$ has generated some trajectory $x(t)=x(t|x_0,u(\cdot))$ (realization of a priori movement) corresponding to initial state x_0.

The function $x(t)$, $t\in T$, after entering the measuring device (2.8) or (2.9) together with $\xi(t)$, $t\in T$, will generate output signals

$$y(t) = Kx(t)$$

or

$$y(t) = Kx(t) + \xi, \ t\in T.$$

The totality $\hat{X}_0$ of all initial states $x_0 \in \check{X}_0$ which together with the chosen control $u(t)$, $t\in T$, and some function $\xi(t)$, $t\in T$ (2.10) are able to generate signal $y(t)$, $t\in T$, will

---

1)* In this case the initial state $x_0 = P^{-1}y(0)$ is restored instantaneously.

be called a posteriori distribution of initial states. This distribution corresponds to the a posteriori movement

$$\hat{X}(t)=\hat{X}(t|u(\cdot))=\{x(t|x_o,u(\cdot)),\ x_o \in \hat{X}_o\},\ t\in T,$$

a posteriori admissible control $\hat{u}(t)$, $t\in T$, $(\hat{X}(t^*|\hat{u}(\cdot)) \subset X^*)$, and a posteriori optimal control $\hat{u}^o(t)$, $t\in T$,

$$(\ J(\hat{u}^o)= \max_{\hat{u}(\cdot)} \min_{z\in\hat{X}(t^*|\hat{u}(\cdot))} h_o'z\ ).$$

As $\hat{X}_o \subset \check{X}_o$, then $J(\hat{u}^o) \geq J(\check{u}^o)$, i.e. using (2.8), (2.9) in the control problem (2.1)-(2.7) we can only increase the efficiency of control.

The purpose of the following sections is to show the concrete method of solving the problem.

2.2.2. Construction of a priori optimal control.

Let $u(\cdot) = (u(t),\ t \in T)$ be some piecewise continuous function satisfying constraint (2.2). According to problem (2.1)-(2.7) where observation for control process is not used we calculate the following numerical characteristics (estimates) of a priori distribution of terminal states:

$$\check{\alpha}_i = \min_{x_o\in\check{X}_o} h_i'x(t^*|x_o,u(\cdot))\ ,\ i = \overline{0,m}. \tag{2.11}$$

Using the Cauchy formula we get

$$X(t^*|x_o,u(\cdot)) = F(t^*)x_o + \int_0^{t^*} F(t^*)F^{-1}(t)bu(t)dt \tag{2.12}$$

where the $n \times n$ matrix function $F(t)$, $t \in T$ is a solution of the equation

$$\dot{F} = AF , \quad F(0)=E,$$

(E is the $n \times n$ unit diagonal matrix).

Substituting (2.12) into (2.11) we obtain

$$\check{\alpha}_i = \int_0^{t^*} h_i' F(t^*) F^{-1}(t) bu(t) dt + \min_{x_0 \in \check{X}_0} h_i' F(t^*) x_0 .$$

Denote

$$p_i'(t) = h_i' F(t), \quad \check{\gamma}_i = \min p_i'(t^*) z , \quad z \in \check{X}_0 .$$

The function $p_i(t)$, $t \in T$, is a solution of equation

$$\dot{p} = Ap \tag{2.13}$$

with initial condition

$$p_i(0)=h_i .$$

Thus the estimates (2.11) equal

$$\check{\alpha}_i = \check{\gamma}_i + \int_0^{t^*} p_i'(t^*) F^{-1}(t) bu(t) dt , \quad i = \overline{0,m}, \tag{2.14}$$

and their calculation is reduced to solving the linear programming problems

$$\check{\gamma}_i = \min p_i'(t^*) z, \tag{2.15}$$

$$Gz = f , \quad d_* \le z \le d^* , \quad i=\overline{0,m}.$$

Using (2.14) we write conditions for a priori control admissibility (2.5)

$$\int_0^{t^*} p_i'(t^*)F^{-1}(t)b\check{u}(t)dt \geq \check{g}_i, \quad i = \overline{1,m} \tag{2.16}$$

where $\check{g}_i = g_i - \gamma_i$.

Similarly, we calculate the value of the quality criterion on the a priori admissible control $\check{u}(t)$, $t\in T$:

$$J(\check{u}) = \check{\alpha}_0 = \min_{x_0\in X_0} h_i'x(t^*|x_0,\check{u}(\cdot)) = \check{\gamma}_0 + \int_0^{t^*} p_0'(t^*)F^{-1}(t)b\check{u}(t)dt. \tag{2.17}$$

According to (2.7),(2.16) the a priori optimal control $\check{u}^0(t)$, $t\in T$ is a solution of the problem

$$\int_0^{t^*} p_0'(t^*)F^{-1}(t)bu(t)dt \longrightarrow \max,$$
$$\int_0^{t^*} p_i'(t^*)F^{-1}(t)bu(t)dt \geq \check{g}_i, \quad i=\overline{1,m}, \quad |u(t)| \leq 1, \quad t\in T. \tag{2.18}$$

In dynamic statement it is the problem

$$J_0(u) = h_0'x(t^*) \longrightarrow \max , \ \dot{x} = Ax + bu, \tag{2.19}$$

$$x(0) = 0, \ h_i'x(t^*) \geq \check{g}_i, \ i=\overline{1,m}, \ |u(t)| \leq 1, \ t \in T.$$

It will be called a determined problem of optimal control accompanying the optimal control problem (2.1)-(2.7) without use of observation.

Finite methods of solving problem (2.19) are described in [19,21].

Thus to construct the a priori optimal control $\check{u}^0(t)$, $t \in T$, of problem (2.1)-(2.7) we need to solve:

1. $(m + 1)$ linear programming problems;
2. one determined problem of optimal control (2.19).

The value of the quality criterion on $\check{u}^0(\cdot)$ equals

$$J(\check{u}^0) = \check{\gamma}_0 + J_0(\check{u}^0)$$

2.2.3. <u>Use of incomplete and inexact measurements.</u>

As a rule,in practice, measurements are conducted with errors. Therefore we consider the most interesting case when control problem (2.1)-(2.7) is connected to the measuring device (2.9)-(2.10).

<u>Remark 2.1.</u> In dynamic system optimization using incomplete and exact measurements (2.8) under general assumptions, the a posteriori distribution of initial states is degenerated at the point $\hat{X}_0 = \{ x_0^0 \}$ and problem (2.1)-(2.7) is reduced to the following:

$$h_0'x(t^*) \longrightarrow \max , \ \dot{x} = Ax + bu, \tag{2.20}$$

$$x(0) = x_0^0, \ h_i'x(t^*) \geq g_i, \ i=\overline{1,m}, \ |u(t)| \leq 1, \ t \in T.$$

Its solution is said to be the ideal optimal control and is denoted by $\overset{\circ}{u}{}^{0}(t), t\in T$.

Let $u(t), t\in T$ be some control under constraints (2.1), $x(t|x_0, u(\cdot))$, $t\in T$ be a trajectory of the system (2.1) generated by this control and an unknown initial state $x_0 \in X_0$, $y(t)$, $t \in T$ be an observable signal of (2.9) generated by the trajectory $x(t|x_0, u(\cdot)), t\in T$ and an unknown function $\xi(t), t\in T$,

$$z(t) = y(t) - \int_0^t KF(t)F^{-1}(\tau)bu(\tau)d\tau, \quad t\in T.$$

For the a posteriori distribution of terminal states $\hat{X}(t^*)$ corresponding to this information we calculate the following estimates

$$\hat{\alpha}_i = \min_{x_0\in\hat{X}_0} h_i' x(t^* | x_0, u(\cdot)) \;, i = \overline{0,m}. \tag{2.21}$$

From (2.21) by using the Cauchy formula we get

$$\hat{\alpha}_i = \hat{\gamma}_i + \int_0^{t^*} p_i'(t^*)F^{-1}(t)bu(t)dt \;, \; i = \overline{0,m}, \tag{2.22}$$

$$\hat{\gamma}_i = \min p_i'(t^*)z \;, \; z \in \hat{X}_0. \tag{2.23}$$

In detail, problems (2.22),(2.23) take the form

$$\hat{\gamma}_i = \min p_i'(t^*)z \;, \tag{2.24}$$

$$\xi_* \le z(t) - K(t)z \le \xi^*, \; t \in T, \tag{2.25}$$

$$Gz = f \;, \; d_* \le z \le d^* \;, \; i=\overline{0,m}. \tag{2.26}$$

Here the $l\times m$ matrix function $K(t)$, $t\in T$, is a solution of the equation

$$\dot{K} = KA, \; K(0) = E \;.$$

Problems (2.24)-(2.26) will be called problems of observation accompanying problem (2.1)-(2.7) with the measuring device (2.9).

A finite algorithm for solving linear semi-infinite problems (2.24)-(2.26) is described in [19].

Having received the estimates $\hat{\gamma}_i, i=\overline{1,m}$, we write the conditions of a posteriori admissibility of control $\hat{u}(t)$, $t\in T$:

$$\int_0^{t^*} p_i'(t^*)F^{-1}(t)bu(t)dt \geq \hat{g}_i, \quad i=\overline{1,m}.$$

Here $\hat{g}_i = g_i - \hat{\gamma}_i$.

The a posteriori optimal control $\hat{u}^0(t), t\in T$ is a solution of problem

$$\int_0^{t^*} p_0'(t^*)F^{-1}(t)bu(t)dt \longrightarrow \max,$$

$$\int_0^{t^*} p_i'(t^*)F^{-1}(t)bu(t)dt \geq \hat{g}_i, \quad i=\overline{1,m}, \quad |u(t)| \leq 1, \quad t\in T. \tag{2.27}$$

In the dynamic statement, problem (2.27) has the form

$$J_0(u) = h_0'x(t^*) \longrightarrow \max, \quad \dot{x} = Ax + bu,$$

$$x(0) = 0, \quad h_i'x(t^*) \geq \hat{g}_i, \quad i=\overline{1,m}, \quad |u(t)| \leq 1, \quad t\in T. \tag{2.28}$$

It will be called a determined problem of optimal control accompanying problem (2.1)-(2.7) with the measuring device (2.9). Problem (2.28) does not differ principally from problem (2.20) and can be solved by the same methods [19].

The value of quality criterion on a posteriori optimal control $\hat{u}^0(t)$ equals

$$J(\hat{u}^o) = J_o(\hat{u}^o) + \hat{\gamma}_o .$$

The value $J(\hat{u}^o) - J_o(\hat{u}^o)$ characterizes the increase of control efficiency throgh the use the measuring device (2.9).

If the same signal $y(t), t\in T$, is observed in (2.9), (2.10) then the value $J(\mathring{u}^o) - J(\hat{u}^o)$ characterizes the decrease in control efficiency because of measurement errors.

It is clear that the optimal control problem (2.1)-(2.7) in which observation (2.9) is used ,is reduced to solving the linear programming problem (2.24)-(2.26) and the determined optimal control problem (2.28). In other words, the problems of observation and control are separated.

Example 2.1. Consider the problem of speed-up of a mass point on horizontal path under the condition that its initial speed is unknown and there is constraint on its terminal state.

Mathematical model of the problem is

$$x_2(1) \longrightarrow \max , \ \dot{x}_1 = x_2, \ \dot{x}_2 = u , x_1(0) = 0, \ |x_2(0)| \le 1, \tag{2.29}$$
$$x_1(0) \le 0.5; \ |u(t)| \le 1, \ t \in T = [0,1].$$

This model is obtained from (2.1)-(2.7) with $n = 2, t^* = 1$,

$$h_0 = \begin{bmatrix} 0 \\ 1 \end{bmatrix}, \ A = \begin{bmatrix} 0 & 1 \\ 0 & 0 \end{bmatrix}, b = \begin{bmatrix} 0 \\ 1 \end{bmatrix}, \ G = 0, \ f = 0, \ d_{1*} = d_1^* = 0,$$

$$d_{2*} = -1, \ d_2^* = 1, \ m = 1, \ h_1 = (-1,0), \ g_1 = -0.5.$$

Let us first construct the a priori optimal control. As in the given example the fundamental matrix of solutions $F(t)$, $t\in [0,1]$ equals

$$F(t) = \begin{bmatrix} 0 & t \\ 0 & 0 \end{bmatrix},$$

then

$$x_1(t) = tx_{10} + \int_0^t (t-\tau)u(\tau)d\tau$$

$$x_2(t) = tx_{20} + \int_0^t u(\tau)d\tau \quad .$$

According to (2.14) the a priori estimate $\check{\alpha}_1$ equals

$$\check{\alpha}_1 = \check{\gamma}_1 + \int_0^1 (1-\tau)u(\tau)d\tau, \; \check{\gamma}_1 = \min_{|z|\le 1} (-z) = -1.$$

Hence $\check{g}_1 = -0.5 + 1 = 0.5$ and the a priori optimal control $\check{u}^0(t)$ is a solution of the determined problem accompanying (2.19):

$$x_2(1) \longrightarrow \max, \; \dot{x}_1 = x_2, \dot{x}_2 = u,$$

$$x_1(0) = x_2(0) = 0, \; x_1(1) \le -0.5,$$

$$|u(t)| \le 1, \; t\in T.$$

The solution to this problem is $\check{u}^0(t) = -1, t\in T$.

The value of the quality criterion $J_0(u)$ equals $J_0(\check{u}^0) = -1$

As the estimate

$$\check{\alpha}_0 = \check{\gamma}_0 + J_0(u) \text{ and } \check{\gamma}_0 = \min_{|z|\le 1} z = -1,$$

the efficiency of a priori optimal control takes the value $J(\check{u}^0) = -2$.

We add the measuring device

$$y = x_1 + x_2 \ , \ (l = 1, \ K=(1,1)). \qquad (2.30)$$

Assume that the output signal is

$$y(t)=v(t), \ v(t) = \int_0^t (t-\tau)u(\tau)d\tau \ (z(t)=y(t)-v(t)=0, \ t\in T).$$

As $x_1(0) = 0$, to restore an unknown initial speed $x_2(0)$, it is enough to measure the signal $y(t)$, $t \in T$, at one moment.

We take $t_1 = 0.5$. From the equation

$$z(0.5)=0.5x_{20} + x_{20}=0$$

we obtain $x_2(0) = 0$. Thus the state $x(0) = (0,0)$ is ideally optimal. The control $\mathring{u}^o(t)$, $t\in T$ is a solution to the determined problem (2.20):

$$x_2(1) \longrightarrow max, \ \dot{x}_1 = x_2, \dot{x}_2=u,$$
$$x_1(0) = x_2(0) = 0, \ x_1(1)\le 0.5, \qquad (2.31)$$
$$|u(t)| \le 1, \ t\in T.$$

and $\mathring{u}^o(t)=1$, $t\in T$. The value of the criterion on this control equals $J(\mathring{u}^o)=1$. Comparing this result with the previous one we conclude that the loss of control efficiency due to the undetermined initial speed equals $J(\mathring{u}^o) - J(\check{u}^o) = 3$.

Now consider the following measuring device:

$$y = x_1 + x_2 + \xi, \qquad (2.32)$$

functioning with errors $\xi(t)$, $t \in T$, which satisfy the equations

$$-0.2 \le \xi(t) \le 0.1, \ t \in T = [0,1].$$

We assume that the signal $y(t) = v(t)$, $t \in T$, has been rewritten, i.e. $z(t) = 0$, $t \in T$. In this case the accompanying observation problems (2.24)-(2.26) are

$$\hat{\gamma}_1 = \max z \ , \ -0.2 \le (1+t)z \le 0.1; \ t \in [0,1], \ -1 \le z \le 1,$$

$$\hat{\gamma}_0 = \min z \ , \ -0.2 \le (1+t)z \le 0.1; \ t \in [0,1], \ -1 \le z \le 1.$$

Hence $\hat{\gamma}_1 = 0.05$, $\hat{\gamma}_0 = -0.1$.

The a posteriori optimal control $\hat{u}^o(t)$, $t \in T$, is the solution of the determined problem

$$x_2(1) \longrightarrow \max, \ \dot{x}_1 = x_2, \ \dot{x}_2 = u, \ x_1(0) = x_2(0) = 0,$$

$$x_1(1) \le 0.45, \ |u(t)| \le 1, \ t \in [0,1],$$

and it is

$$\hat{u}^o(t) = -1, \ t \in [0, \tau_1], \ \hat{u}^o(t) = 1, \ t \in [\tau_1, 1],$$

$$\tau_1 = 1- \qquad 0.95 = 0.02532.$$

The criterion takes the value

$$J(\hat{u}^o) = 2 \times 0.95 - 1.1 = 0.8493.$$

The increase in control efficiency due to observation by (2.32) equals

$$J(\hat{u}^o) - J(\check{u}^o) = 2 \times 0.95 + \ 0.9 \ = 2.8493.$$

The loss of control efficiency because of measurement inaccuracy (2.32) equals

$$J(\mathring{u}^o) - J(\hat{u}^o) = 2.1 - 2 \times 0.95 = 0.1506.$$

2.3. OPTIMIZATION OF DYNAMIC SYSTEMS WITH IDENTIFICATION OF INPUT PERTURBATIONS.

Every real control system is functioning, as a rule, in the presence of noise. This leads to the necessity of introduction of uncertainties into the mathematical equation which describe the behaviour of dynamic objects.

In this section we shall construct a finite algorithm for optimal control by dynamic systems with perturbations.

2.3.1. Optimization problem of dynamic systems with perturbations.

We consider the family of q - vector functions $\omega(t)$, $t\in T = [0, t^*]$:

$$\omega(t) = \omega_0(t) + \sum_{i=1}^{q} w_i\omega_i(t) , \tag{3.1}$$

defined by the fixed piecewise continuous p-vector function $\omega_0(t)$, $\omega_1(t)$, ..., $\omega_q(t)$, $t\in T$, and the q-vector of parameters $w= = (w_1, \ldots, w_q)$, which may take any value in the set

$$\check{W} = \{ w\in R^q : Gw = f, \ d_* \le w \le d^* \} \ (f\in R^l) . \tag{3.2}$$

We shall assume that function (3.1) describes perturbations acting upon the dynamic system

$$\dot{x}(t) = A(t)x + b(t)u + D(t)\omega(t), \ x(0) = x_0 \ (x\in R^n, \ u\in R) \tag{3.3}$$

with piecewise continuous elements $A(t)$, $b(t)$, $D(t)$, $t\in T$.

To every piecewise control $u(\cdot) = (u(t), \ t\in T)$ the only corresponding movement is

$$\check{X}(t) = \check{X}(t|x_0, u(\cdot)) = \{ x(t|x_0, u(\cdot), w), w\in\check{W} \}, t\in T,$$

consisting of all trajectories $x(t) = x(t|x_0, u(\cdot), w)$, $t\in T$, of system (3.3) generated by the fixed initial state x_0, the control $u(\cdot)$ and different parameters $w\in\check{W}$.

In the following we shall call $\check{W}$ the a priori distribution of parameters and $\check{X}(t)$, $t\in T$, the a priori movement of system (3.3).

Let the terminal set X^* (2.4) in the state space of system (3.3) be given.

The control $\check{u}(t)$, $t\in T$, constrained by (2.1) and the corresponding motion $\check{X}(t|\check{u}(\cdot))$, $t \in T$, will be called a priori admissible if the terminal inclusion (2.5) is true.

The quality of the a priori admissible control $u^o(\cdot)$ will be estimated according to the functional value

$$J(\check{u}) = \min_{w\in\check{W}} h_0' x(t^*|x_0, \check{u}(\cdot), w). \tag{3.4}$$

As usually the control $\check{u}^o(\cdot)$ will be called a priori optimal if

$$J(\check{u}) = \max_{\check{u}(\cdot)} J(\check{u}). \tag{3.5}$$

By analogy to Section 2.2 we shall introduce the procedure of observation over the control process.

Consider the following types of linear inertia-free measuring systems:

1) direct complete,

$$y = Cw \quad (y\in R^q, \ det\ C \neq 0) ; \tag{3.6}$$

2) direct incomplete exact,

$$y = Cw \quad (y\in R^l, \ \mathrm{rank}\ C = l < q) ;$$

3) indirect incomplete exact,

$$y = Kx \quad (y \in R^l) \; ; \tag{3.7}$$

4) mixed inexact incomplete,

$$y = Kx + Cw + \xi . \tag{3.8}$$

In case (3.8) we shall assume that any piecewise continuous function $\xi(t)$, $t \in T$, satisfying inequalities (2.10) may be implemented.

Let for the chosen control $u^*(\cdot)$ the measuring device have a registered signal $y^*(t)$, $t \in T$.

The set $\hat{W}$ of vectors, $w \in \check{W}$, which together with some possible error function $\xi(t)$, $t \in T$, is able to generate the signal $y^*(t), t \in T$, will be called the a posteriori distribution of the parameters w. The a posteriori motion

$$\hat{X}(t|x_0, u^*(\cdot)) = \{ x(t|x_0, u^*(\cdot), w) \; w \in \hat{W} \}, \; t \in T,$$

the a posteriori admissible control

$$\hat{u}^*(t), \; t \in T, \; (\hat{X}(t^* | x_0, \hat{u}^*(\cdot)) \subset X^*)$$

and the a posteriori optimal control

$$\hat{u}^o(\cdot) \; (J(\hat{u}^o) = \max_{\hat{u}^*(\cdot)} J(\hat{u}^*))$$

correspond to it.

Now we shall construct a priori optimal controls.

Consider the control problem (3.1)-(3.5) without the use of observation over the control process.

In accordance with control problem (3.1)-(3.5) we calculate the following estimates of the a priori distribution $\check{W}$:

$$\check{\alpha}_i = \min_{w \in \check{W}} h_i' x(t^* | x_0, u(\cdot), w), \; i = \overline{0, m}. \tag{3.9}$$

Using the Cauchy formula we get

$$x(t^*|x_0,u(\cdot),\ w) = F(t^*)x_0 + \int_0^{t^*} F(t^*)F^{-1}(t)b(t)u(t)dt +$$

$$+ \int_0^{t^*} F(t)F(t)D(t)\omega\ (t)dt + \sum_{j=1}^{q} w_j \int_0^{t^*} F(t^*)F^{-1}(t)D(t)\omega_j(t)dt.$$

Using (3.9) we obtain

$$\check{\alpha}_i = \check{\gamma}_i + h_i'F(t^*)x_0 + \int_0^{t^*} h_i'F(t^*)F^{-1}(t)b(t)u(t)dt +$$

$$+ \int_0^{t^*} h_i'F(t^*)F^{-1}(t)D(t)\omega_0(t)dt, \qquad (3.10)$$

$$\check{\gamma}_i = min\ a_i'w,\quad Gw = f,\quad d_* \le w \le d^*,\ i = \overline{0,m}$$

$$a_i = (a_{ij},\ j = \overline{1,\ q}),\quad a_{ij} = \int_0^{t^*} F(t^*)F^{-1}(t)D(t)\omega_j(t)dt.$$

With the help of estimates (3.9) the conditions of the a priori admissibility of control $\check{u}(\cdot)$ can be written in the form

$$h_i'F(t^*)x_0 + h_i' \int_0^{t^*} F(t^*)F^{-1}(t)b(t)u(t)dt +$$

$$+ h_i' \int_0^{t^*} F(t^*)F^{-1}(t)D(t)\omega_0(t)dt \ge \check{g}_i,\ i = \overline{1,\ m}.$$

Find the value of the quality criterion on the a priori admissible control $\check{u}(t)$, $t\in T$,

$$\check{J}(u) = \check{\alpha}_0 = \check{\gamma}_0 + h_0'F(t^*)x_0 + \int_0^{t^*} h_0'F(t^*)F^{-1}(t)b(t)u(t)dt +$$

$$+ \int_0^{t^*} h_0'F(t^*)F^{-1}(t)D(t)\omega_0(t)dt.$$

According to (3.5) the a priori optimal control $\check{u}^0(t)$, $t \in T$, is solution of the problem

$$h_0'F(t^*)x_0 + \int_0^{t^*} h_0'F(t^*)F^{-1}(t)b(t)u(t)dt +$$

$$+h_0'\int_0^{t^*} F(t^*)F^{-1}(t)D(t)\omega_0(t)dt \longrightarrow max,$$

$$h_i'F(t^*)x_0 + h_i'\int_0^{t^*} F(t^*)F^{-1}(t)b(t)u(t)dt +$$

$$+ h_i'\int_0^{t^*} F(t^*)F^{-1}(t)D(t)\omega_0(t)dt \geq \check{g}_i, \quad i = \overline{1, m};$$

$$|u(t)| \leq 1, \ t \in T.$$

In the dynamic statement it has the form

$$\begin{gathered} J_0(u) = h_0'x(t^*) \longrightarrow max \\ \dot{x} = A(t)x + b(t)u(t) + D(t)\omega_0(t); \ x(0) = x_0 \\ h_i'x(t^*) \geq \check{g}_i, \quad i = \overline{1,m}; \ |u(t)| \leq 1, \ t \in T. \end{gathered} \tag{3.11}$$

Problem (3.11) is a determined optimal control problem accompanying problem (3.1)-(3.5) for constructing the a priori optimal control.

So, to construct an a priori optimal control $\check{u}^{o}(t)$, $t \in T$, for (3.1)-(3.5) we need to solve:

1) $(m + 1)$ - problems of linear programming (3.10),
2) one optimal control problem (3.11).

The value of the quality criterion on $\check{u}^{o}(\cdot)$ equals

$$J(\check{u}^{o}) = \gamma_{o} + J_{o}(\check{u}^{o}).$$

2.3.2. Optimization of perturbed dynamic control systems by means of observation .

Let us add to the control procedure the operations on treatment of output signals of the measuring device (3.7) or (3.8). In Section 2.3.1 the notion was introduced that the a posteriori distribution $\hat{W}$ of parameters in general (i.e. nonconstructive) form contains the complete results of the removed uncertain elements from the set $\check{W}$ by means of the observed signal $y(t)$, $t \in T$. To solve the problem it is sufficient to introduce only separate numerical characteristics (estimates) of the set $\hat{W}$.

We calculate

$$\hat{\alpha}_{i} = \min_{w \in \hat{W}} h_{i}' x(t^{*} | x_{o}, u^{*}(\cdot), w), \quad i = \overline{0, m}. \qquad (3.12)$$

(3.12) will be called identification problems accompanying the optimization problem under uncertainty conditions.

Consider the case of indirect incomplete exact measurement (3.7). Let the control $u^{*}(t)$, $t \in T$, be given on input of system (3.3). It generates the trajectory $x(t | x_{o}$, $u^{*}(\cdot)$, $\omega(\cdot))$, $t \in T$ (x_{o}, $\omega(\cdot)$, $t \in T$, are given) with some value of the parameter $w \in \check{W}$. The measuring device (3.7) yields the necessary signal $y^*(t)$, $t \in T$. Under conditions [15], moments $t_j \in T$, $j = \overline{1,p}$, and sets L_j, $|L_j| \le 1$, $j = \overline{1,p}$; $\sum_{j=1}^{p} |L_j| = q$, will be found such that the matrix

$$P = \begin{pmatrix} K(L_j)F(t_j) \\ j = \overline{1,p} \end{pmatrix}$$

be non-singular.

Here $K(L_j)$ is a submatrix of K containing rows with indices from L_j. Now we form the signal

$$v(t) = \int_0^t KF(t)F^{-1}(\tau)b(\tau)u(\tau)d\tau + KF(t)x_0 +$$

$$+\int_0^t KF(t)F^{-1}(\tau)D(\tau)\omega_0(\tau)d\tau$$

Assume

$$y^*(t) - v(t) = z(t).$$

Compose the n-vector $z_{sup} = (z_s(t_s),\ s \in L_j,\ j = \overline{1,p})$ and find the unknown parameter $w^0 = P^{-1}z_{sup}$.

Therefore, the a posteriori parameters' distribution generates a fixed element w^0 (uncertainty of problem (3.1)-(3.5) disappears) for the exact measurements (3.7) of the support signals $y^*_s(t)$, $s \in L_j$, $t \in T$, at the support moments t_j, $j = \overline{1,p}$. After constructing w^0 problem (3.1)-(3.5),(3.7) becomes

$$h_0'x(t^*) \longrightarrow \max,\quad \dot{x} = A(t)x + b(t)u + D(t)\omega_0(t),\ x(0) = x_0,$$

$$h_i'x(t^*) \ge \hat{g}_i,\ i = \overline{1,m}\ ;\ |u(t)| \le 1,\ t \in T. \qquad (3.13)$$

It will be called a determined problem of optimal control accompanying (3.1)-(3.5) with the measuring device (3.7).

The solution $\mathring{u}^o(t)$, $t\in T$ of problem (3.16) is an ideal optimal control.

The value $J(\mathring{u}^o) - J(\check{u}^o)$ characterizes the increase of control efficiency by using the measuring device (3.7).

Consider the most interesting case when problem (3.1)-(3.5) is connected with the measuring device (3.8).

Let $u^*(t)$, $t \in T$, be some control with constraints (3.4), $x(t|x_0, u^*(\cdot), w)$, $t \in T$, some system trajectory generated by this control, the given initial state x_0 and the known value of parameter $w\in\check{W}$; let $y^*(t)$, $t \in T$, be an observed signal of the measuring device (3.8) caused by trajectory $x(t|x_0, u^*(\cdot), w)$, $t\in T$, and an unknown function $\xi(t)$, $t\in T$. Assume

$$z(t) = y^*(t) - \int_0^t KF(t)F^{-1}(\tau)b(\tau)u^*(\tau)d\tau - KF(t)x_0 -$$

$$- \int_0^t KF(t)F^{-1}(\tau)D(\tau)\omega_0(\tau)d\tau.$$

For the a posteriori distribution of the terminal states $\hat{X}(t^*)$ we calculate the estimates

$$\hat{\alpha}_i = \min_{w\in\hat{W}} h_i'x(t^*|x_0,u^*(\cdot),w), \quad i = \overline{0, m}. \tag{3.14}$$

Using Cauchy's formula from (3.14) we obtain

$$\hat{\alpha}_i = \hat{\gamma}_i + h_i'F(t^*)x_0 + \int_0^{t^*} h_i'F(t^*)F^{-1}(t)b(t)u^*(t)dt +$$

$$+ \int_0^{t^*} h_i'F(t^*)F^{-1}(t)D(t)\omega_0(t)dt, \quad i = \overline{0, m};$$

$$\hat{\gamma}_i = \min a_i'w, \ w \in \hat{W}.$$

Calculation of numbers $\hat{\gamma}_i$, $i = \overline{0,m}$, in a detailed notation gives the problems

$$\hat{\gamma}_i = \min a_i'w, \quad Gw = f, \ d_* \le w \le d^*, \ i = \overline{0, m},$$

$$\xi_* \le z(t) - Kw'd \le \xi^*, \tag{3.15}$$

$$d = (d_j, \ j = 1, q), \ d_j = \int_0^{t^*} F(t^*)F^{-1}(t)D(t)\omega_j(t)dt.$$

Problem (3.15) is said to be the identification problem accompanying the optimal control problem (3.1)-(3.5) with device (3.8).

Having given the estimates $\hat{\gamma}_i$, $i = \overline{1,m}$,we write the conditions for the a posteriori admissibility of control $u(t)$, $t \in T$,

$$h_i'F(t^*)x_0 + h_i' \int_0^{t^*} F(t^*)F^{-1}(t)b(t)u(t)dt +$$

$$+ h_i' \int_0^{t^*} F(t^*)F^{-1}(t)D(t)\omega_0(t)dt \ge \hat{g}_i, \ i = \overline{1, m}$$

where $\hat{g}_i = g_i - \hat{\gamma}_i$, $i = \overline{1, m}$.

The a posteriori optimal control $\hat{u}^0(t)$, $t \in T$, is the solution to the problem

$$J_0(u) = h_0'F(t^*)x_0 + \int_0^{t^*} h_0'F(t^*)F^{-1}(t)b(t)u(t)dt +$$

$$+h'_0\int_0^{t^*} F(t^*)F^{-1}(t)D(t)\omega_0(t)dt \longrightarrow max,$$

$$h'_iF(t^*)x_0 + h'_i\int_0^{t^*} F(t^*)F^{-1}(t)b(t)u(t)dt +$$

$$+ h'_i\int_0^{t^*} F(t^*)F^{-1}(t)D(t)\omega_0(t)dt \geq \hat{g}_i, \quad i = \overline{1, m},$$

$$|u(t)| \leq 1, \quad t\in T. \tag{3.16}$$

The dynamic form problem (3.16) can be written

$$J_0(u) = h'_0x(t^*) \longrightarrow max , \quad \dot{x} = A(t)x + b(t)u + D(t)\omega_0(t),$$

$$x(0) = x_0, \quad h'_ix(t^*) \geq \hat{g}_i, \quad i = \overline{1,m}; \quad |u(t)| \leq 1, \quad t\in T. \tag{3.17}$$

Problem (3.17) is a determined optimal control problem, accompanying (3.1)-(3.5) with device (3.8).

The value of the quality criterion for the a posteriori optimal control $\hat{u}^0(t)$, $t\in T$, is equal to $J(\hat{u}^0) + \hat{\gamma}_0$. The value $J(\hat{u}^0) - J(\check{u}^0)$ characterizes the increase of control efficiency at the expense of observation results by means of the measuring device (3.8). The number $J(\mathring{u}^0) - J(\hat{u}^0)$ is equal to the loss of control efficiency because of measurement errors.

From (3.16) it can be seen that the identification problem does not depend on control, i.e. the problems of control and identification are separate.

<u>Example 3.1.</u> Consider the problem of the acceleration of a mass point on horizontal section of path affected by unknown constant force.

The mathematical simulation of the problem is as follows:

$$x_2(1) \longrightarrow \max, \ \dot{x}_1 = x_2, \ \dot{x}_2 = u + w, \ x_1(0) = x_2(0) = 0,$$

$$x_1(1) \le 0.5; \ |u(t)| \le 1, \ t \in T = [0, 1], \ 0 \le w \le 1.$$

It presents a special case of the problem (3.3) :

$$n = 2, \ h_0 = (0, 1), \ A = \begin{pmatrix} 0 & 1 \\ 0 & 0 \end{pmatrix}, \ b = \begin{pmatrix} 0 \\ 1 \end{pmatrix},$$
$$w \in \check{W} = \{ w \in R^1 : 0 \le w \le 1 \};$$

$$\check{\Omega}(\cdot) = \{ \omega(t), \ t \in T, \ \omega(t) \equiv w \}, \ g = 1, \ \omega_0(t) \equiv 0,$$

$$\omega_1(t) \equiv 1, \ t \in T; \quad p = 1, \ m = 1, \ h_1 = (-1, 0), \ g_1 = -0.5.$$

First we construct the a priori optimal control. Since the fundamental matrix of solutions $F(t)$, $t \in T$, takes the form $F(t) = \begin{pmatrix} 1 & t \\ 0 & 1 \end{pmatrix}$,

$$x_1(t) = \int_0^t (t - \tau)u(\tau)d\tau + wt^2/2, \quad x_2(t) = \int_0^t u(\tau)d\tau + wt.$$

According to (3.9) the a priori estimates $\check{\alpha}_1$, $\check{\alpha}_0$ are equal to

$$\check{\alpha}_1 = \int_0^1 (1 - t)u(t)dt + \check{\gamma}_1, \qquad \check{\alpha}_0 = \int_0^1 u(t)dt + \check{\gamma}_0,$$

$$\check{\gamma}_1 = \max_{0 \le w \le 1} w/2 = 0.5, \qquad \check{\gamma}_0 = \min_{0 \le w \le 1} w = 0.0$$

Hence, $\check{g}_1 = -0.5 + 0.5 = 0$ and an a priori optimal control $\check{u}^0(t)$, $t \in T$, is the solution to the accompanying determined problem

$$x_2(1) \longrightarrow \max, \quad \dot{x}_1 = x_2, \; \dot{x}_2 = u, \quad x_1(0) = x_2(0) = 0,$$

$$x_1(1) \le 0, \; |u(t)| \le 1, \; t \in T.$$

The solution to the problem has the form

$$\check{u}^0(t) = -1, \; t \in [0, \; 1 - \sqrt{2}/2], \; \check{u}^0(t) = 1, \; t \in [1 - \sqrt{2}/2, \; 1].$$

The value of the quality criterion is equal to

$$J(\check{u}^0) = \sqrt{2} - 1 \approx 0.414.$$

Assume that the signal

$$y^*(t) = \int_0^t (t - \tau)u(\tau)d\tau, \; t \in [0, \; 1],$$

is given by the measuring device

$$y = x_1 + x_2.$$

Then $\mathring{w} = 0$ and the control $\mathring{u}^0$, $t \in T$, is the solution to the determined problem

$$x_2(1) \longrightarrow \max, \quad \dot{x}_1 = x_2, \; \dot{x}_2 = u, \quad x_1(0) = x_2(0) = 0,$$

$$x_1(1) \le 0.5, \quad |u(t)| \le 1, \; t \in T,$$

and has the form $\mathring{u}^0(t) = 1$, $t \in T$. The value of quality criterion is equal to $J(\mathring{u}^0) = 1$.

Now, consider the measuring device

$$y = x_1 + x_2 + \xi, \tag{3.18}$$

operating with errors $\xi(t)$, $t \in T$, satisfying the restrictions $-0.2 \le \xi(t) \le 0.1$, $t \in [0, \; 1]$. Assume again that the signal

$$y^*(t) = \int_0^t (t - \tau + 1)u(\tau)d\tau, \; t \in T,$$

is given. In this case the accompanying identification problems have the form

$$\hat{\gamma}_1 = \max_{0 \le w \le 1} w/2, \quad -0.2 \le (t^2/2 + t)w \le 0.1, \ t \in [0, 1];$$

$$\hat{\gamma}_0 = \min w, \quad -0.2 \le (t^2/2 + t)w \le 0.1, \ t \in [0, 1]; \ 0 \le w \le 1.$$

Hence, $\hat{\gamma}_1 = 0.0333$, $\hat{\gamma}_0 = 0$. Then $\hat{g}_1 = -0.5 + 0.0333 \approx 0.47$. The a posteriori optimal control $\hat{u}^0(t)$, $t \in T$, is the solution to the accompanying determined problem

$$x_2(1) \longrightarrow \max, \ \dot{x}_1 = x_2, \ \dot{x}_2 = u, \quad x_1(0) = x_2(0) = 0,$$

$$x_1(1) \le 0.47, \quad |u(t)| \le 1, \ t \in [0, 1],$$

and has the form

$$\hat{u}^0(t) = -1, \ t \in [0, \tau_1]; \quad \hat{u}^0(t) = 1, \ t \in [\tau_1, 1],$$

$$\tau_1 = 1 - \sqrt{0.97} \approx 1 - 0.9849 = 0.0151.$$

The quality criterion for it takes the value $J(\hat{u}^0) = 2\sqrt{0.97} - 1 = 0.9698$. The increase in control efficiency at the expense of observation by means of device (3.18) is as follows:

$$J(\hat{u}^0) - J(\check{u}^0) = 0.9698 - 0414 = 0.5558.$$

The control efficiency loss due to inaccuracy of measurements is equal to

$$J(\mathring{u}^0) - J(\hat{u}^0) = 1 - 0.9698 = 0.0302.$$

2.4. OPTIMAL ESTIMATOR FOR DYNAMIC SYSTEMS.

The observation problem is the most important in the theory of control systems under uncertainty conditions.

R. Kalman [24] and his successors created the finely determined theory of observability dual to the theory of controllability.

In Section 2.1 we proposed the new view on the relation between problems of control and observation. Now we shall obtain equations based on it for an optimal estimator for continuous linear systems.

We consider the dynamic system, the behaviour of which on the interval of time $T = [t_*, t^*]$ is described by

$$\dot{x} = A(t)x \tag{4.1}$$

where $A(t)$, $t \in T$, is an $n{\times}n$ piecewise continuous matrix function.

Assume that the exact initial state $x(t_*)$ of system (4.1) is not known. A priori information about it has the form

$$x(t_*) \in \check{X}_* = \{ x \in R^n : Gx = f,\ d_* \le x \le d^* \},\ (f \in R^m). \tag{4.2}$$

The a priori distribution $\check{X}_*$ of initial states generates the a priori motion

$$\check{X}(t) = \{ x(t|z),\ z \in \check{X}_* \},\ t \in T,$$

composed of different trajectories $x(t|z)$, $t \in T$, of system (4.1) outgoing from points $x(t_*) = z \in \check{X}_*$ at the moment t_*.

In control problems the a priori distribution $\check{X}(t^*)$ of terminal states often does not allow us to construct effective control due to excessive uncertainties. To reduce the uncertainty of terminal states we introduce the procedure of observation. Assume that there is the measuring device

$$y = C(t)x + \xi, \quad (y \in R^k),$$

It registers the linear combinations $C(t)x(t)$ components of the state $x(t)$ at each moment $t \in T$ with errors $\xi(t)$.

The matrix function $C(t)$, $t \in T$, is considered to be piecewise continuous. Any piecewise continuous functions $\xi(t)$, $t \in T$, satisfying unequalities

$$\xi_* \le \xi(t) \le \xi^*, \quad t \in T, \; (\xi_*, \xi^* \in R^k) \qquad (4.3)$$

can be realized as errors of measurement.

Let us carry out observation on the interval $T_\theta = [t_*, \theta]$ and register the signal $y(t)$, $t \in T_\theta$. Information obtained from the signal $y(t)$, $t \in T_\theta$, allows us to decrease the uncertainty of system (4.1) at any moment of time.

The set $\hat{X}^\theta_*$ will be called the a posteriori distribution of initial states of the system (4.1) if it is composed of those and only those initial states $z = x(t_*) \in \check{X}_*$, that are able, together with some errors of measurement $\xi(t)$, $t \in T_\theta$, to generate the observed signal $y(t)$, $t \in T_\theta$.

The a posteriori motion $\hat{X}^\theta(t) = x(t|z), z\in\hat{X}^\theta_*\}$, $t \in T_\theta$, and the a posteriori distribution $\hat{X}^\theta(t^*)$ of terminal states of system (4.1) correspond to a posteriori distribution of initial states $\hat{X}^\theta_*$.

As a rule, full information about the sets $\check{X}_*$, $\hat{X}^\theta_*$, $\check{X}(t^*)$, $\hat{X}^\theta(t^*)$, is not used in control problems. In each particular case it is enough to know the certain numerical characteristics (estimates) of these sets. We assume that the linear control problems have to deal with linear estimates of the form

$$\hat{\alpha}^\theta = \max_{z\in\hat{X}^\theta} h'x(t^*|z) \qquad (4.4)$$

Calculation of estimates (4.4) will be called a linear problem of observation.

We introduce the functions

$$h'(t) = h'F(t,t_*), \quad M(t) = C(t)F(t,t_*), \quad t \in T$$

where $F(t, \tau)$, t, $\tau \in T$, is the fundamental matrix of system (4.1):

$$\frac{\partial F}{\partial t} = A(t)F \ , \ F(t_*,t_*) = E$$

Here E is the n×n unit diagonal matrix.

In terms of these functions, problem (4.1)-(4.4) takes the form

$$\hat{\alpha}^{\theta} = \max h'(t^*)z \ , \ \xi_* \le y(t) - M(t)z \le \xi^*, \quad t \in T_\theta \ , \qquad (4.5)$$

$$Gz = f, \quad d_* \le z \le d^* .$$

While constructing optimal feedback controls the estimates of a posteriori distribution of terminal states are to be calculated in the mode of real time. It is clear that to do it directly, solving problem (4.5) for various $\theta \ge t_*$ is unreasonable and practically impossible owing to the excessive demands on computer speed.

We describe the laws of variation solution for problem (4.5) depending on time θ. To simplify calculations we shall limit ourselves to the case $G = 0$, $f = 0$, $k = 1$. According to [19] the optimal support plan $\{z(\theta),S_{sup}(\theta)\}$ is the solution to the simple problem (4.5). The support $S_{sup}(\theta)$ consists of two components $\{\tau_{sup}(\theta),I_{sup}(\theta)\}$. The first component $\tau_{sup}(\theta)$ is the family of time moments

$$t_* \le \tau_1(\theta) \le \tau_2(\theta) \le \ \ldots \ \le \tau_l(\theta) \le \theta.$$

The second component $I_{sup}(\theta)$ contais l indices from the set $I = \{ 1,2,\ldots,n \}$ of indices of the feasible solution z. The $l \times l$ non-singular support matrix

$$M_{sup}(\theta) = M(\tau_{sup}(\theta),I_{sup}(\theta)) = \begin{pmatrix} M_j(\tau_i(\theta)) \ : \ j \in I_{sup} \\ i = \overline{1, l} \end{pmatrix}$$

corresponds to support $S_{sup}(\theta)$, where $M_j(t)$ is the j-th column of matrix $M(t)$.

Support $S_{sup}(\theta)$ is accompanied by the vector of potentials

$$\nu = \nu(\tau_{sup}(\theta)) = (\nu(\tau_1(\theta)), \quad ,\nu(\tau_l(\theta))),$$

$$\nu' = h'_{sup}(t^*)M^{-1}_{sup}(\theta)$$

where

$$h_{sup}(t^*) = (h_j(t^*), \; j \in I_{sup}(\theta)).$$

The vector of potentials ν together with vector of estimates

$$\Delta'(\theta) = \Delta'(\theta|I_N) = (\Delta_j(\theta), j\in I_N)' =$$

$$= \nu' M(\tau_{sup}(\theta), I_N) - h'_N(t^*), \quad I_N = I\backslash I_{sup},$$

composes the basis of optimality criterion [19]:

$$\Delta_j(\theta) \ge 0 \text{ if } z_j(\theta) = d_{*j} \; ; \; \Delta_j(\theta) \le 0 \text{ if } z_j(\theta) = d^*_j \; ;$$

$$\Delta_j(\theta) = 0 \text{ if } d_{*j} \le z_j(\theta) \le d^*_j \; ; \; j \in I_N \; ;$$

$$\nu(\tau_i(\theta)) \le 0 \text{ if } y(\tau_i(\theta)) - M(\tau_i(\theta))z(\theta) = \xi^* \; ;$$

$$\nu(\tau_i(\theta)) \ge 0 \text{ if } y(\tau_i(\theta)) - M(\tau_i(\theta))z(\theta) = \xi_* \; ;$$

$$\nu(\tau_i(\theta)) = 0 \text{ if } \xi_* \le y(\tau_i(\theta)) - M(\tau_i(\theta))z(\theta) \le \xi^* \; ;$$

$$i = \overline{1, l}$$

Let θ be a moment of time such that:

1) The functions $A(t)$, $C(t)$, $y(t)$, $t \in T_\theta$, are continuous together with the second derivatives in the neighbourhood of points $\tau_i(\theta)$, $i = \overline{1,l}$, $y(\tau_i(\theta)) - M(\tau_i(\theta))z(\theta) \ne 0$;

2) $d_{*sup} \le z_{sup}(\theta) \le d^*_{sup}$;

3) $\xi_* \le y(t) - M(t)z(\theta) \le \xi^*$, $t \in T_\theta \backslash \tau_{sup}(\theta)$;

4) $\nu(\tau_{sup}(\theta)) \ne 0$;

5) $\Delta(\theta\ I_N) \neq 0$;

6) $\tau_l(\theta) = \theta$, $\dot{y}(\theta) - \dot{M}(\theta)z(\theta) \neq 0$.

Then in the neighbourhood of point θ the component $I_{sup}(\theta)$ of optimal support $S_{sup}(\theta)$ is constant and the component $\tau_{sup}(\theta)$ and the support components $z_{sup}(\theta)$ of the optimal feasible solution $z(\theta)$ satisfy the equations

$$M_{sup}(\tau_i)\dot{z}_{sup} = 0, \ i = \overline{1,l-1}$$

$$M_{sup}(\theta)\dot{z}_{sup} = \dot{y} - \dot{M}z,$$

$$(y(\tau_i) - M(\tau_i)z)\dot{\tau}_i = \dot{M}_{sup}(\tau_i)\dot{z}_{sup}, \ i = \overline{1, l-1} \ , \tag{4.6}$$

$$\dot{\tau}_i = 1 \ .$$

The correlations (4.6) will be called the equations of the optimal estimator. Initial conditions for $\theta = t_*$ for equations (4.6) are obtained from the a priori distribution.

Equations (4.6) are integrated after entering results of measuring $y(\theta)$. Their form changes together with variation of component $I_{sup}(\theta)$ of optimal support $S_{sup}(\theta)$.

The rules of variation $I_{sup}(\theta)$ are obtained from those of variation of support in the adaptive linear programming method (see the Appendix).

Example 4.1. As an illustration we consider the problem of observation for the motion of material point on a rectilinear section of path under action of constant force. The mathematical model of problem has the form

$$\dot{x} = w \ . \tag{4.7}$$

Suppose that at initial moment $t_* = 0$ the material point was found in neighbourhood of $x = 0$:

$$|\dot{x}(0)| \leq 1.$$

At this moment let its unknown velocity $\dot{x}(0)$ satisfy the inequality

$$|\dot{x}(0)| \le 1.$$

Let it also be known that the force acting at the point can take any value from set

$$\check{W} = \{ w \in R^1 : |w| \le 2 \}.$$

It is necessary to estimate the maximal possible value of the point velocity at the current moment θ.

Evidently, the a priori estimate of velocity equals $\hat{\alpha}^{\theta} = 1+2\theta$. It is clear that for large $\theta > 0$ this estimate may be extremely gross.

We supply system (4.7) with the device

$$y = x + \xi$$

which is able instantly (without inertia) to fix the position of a point with an error ξ

$$|\xi(t)| \le 1, \quad t \in [0, \theta].$$

Let us investigate the situation when the measuring device registers the signal $y(t) \equiv 0$, $t \ge 0$.

If we introduce the phase variables

$$x_1 = x, \quad x_2 = \dot{x}, \quad x_3 = w,$$

the vectors

$$h = (0, 1, 0), \; c = (1, 0, 0),$$

and the matrix

$$A = \begin{pmatrix} 0 & 1 & 0 \\ 0 & 0 & 1 \\ 0 & 0 & 0 \end{pmatrix},$$

then our problem becomes a special case of problem (4.5) when

$$n = 3, \; m = 1, \; G = 0, \; f = 0,$$

$$d_* = (-1, -1, -2), \; d^* = (1, 1, 2),$$

$$\xi_* = -1, \; \xi^* = 1, \; t^* = \theta .$$

The observation problem (4.5) for $\theta = 4$ takes the form

$$x_2 + 8x_3 \longrightarrow max, \quad |x_1 + tx_2 + t^2x_3| \le 1$$
$$t \in [0,\ 4], \ |x_i| \le 1, \ i = \overline{1,3} \tag{4.8}$$

where $x_3 = w/2$.

The solution of problem (4.8) was obtained in [19]:

$$x^0 = (-0.236068;\ -1.0;\ 0.327254);\ \tau_1(4) = 1.527864;$$
$$\tau_2(4) = 4;\quad I_{sup}(4) = \{1,\ 3\}. \tag{4.9}$$

The equation of the optimal estimator for $\theta \in [\sqrt{2}+1/2, 8]$ is

$$I_{sup}(\theta) = \{1,\ 3\};$$
$$\dot{x}_1 + \tau_1\dot{x}_3 = 0;$$
$$x_1 + \theta^2 x_3 = 1 + \theta;$$
$$\dot{\tau}_1 = -\tau_1\dot{x}_3/x_3\ ;$$
$$\tau_2 = \theta$$

with initial conditions (4.9).

The optimal estimator for $\theta \in [8,\ \infty]$ is

$$I_{sup}(\theta) = \{\ 2,\ 3\ \};$$
$$\dot{x}_2 + + \tau_1\dot{x}_3 = 0;$$
$$x_2 + \theta x_3 = 1;$$
$$\dot{\tau}_1 = -(\dot{x}_2 + 2\tau_1\dot{x}_3)/2x_3;$$
$$\tau_2 = \theta.$$

The initial conditions are

$$x_2(8) = -1;\ x_3(8) = 1/8;\ \tau_1(8) = 4;\ \tau_2(8) = 8.$$

In the given case the solution is found in the explicit form

$$x_2(\theta) = -8/\theta;\ x_3(\theta) = 8/\theta^2;\ \tau_1(\theta) = \theta\ .$$

For $\theta \in [\ \sqrt{2},\ \sqrt{2} + 1\ ]$ we have

$$I_{sup}(\theta) = \{ 1, 2 \};$$

$$\dot{x}_1 + \tau_1 \dot{x}_2 = 0; \; x_1 + \theta x_2 = 1 - \theta^2;$$

$$\dot{\tau}_1 = -\dot{x}_2/2; \; \tau_2 = \theta;$$

$$x_1(\sqrt{2} + 1/2) = -3/4; \; x_2(\sqrt{2} + 1/2) = -1;$$

$$\tau_1(\sqrt{2} + 1/2) = 1/2; \; \tau_2(\sqrt{2} + 1/2) = \sqrt{2} + 1/2.$$

For $\theta \in [1, \sqrt{2}]$:

$$I_{sup}(\theta) = \{2\}; \; x_2(\theta) = (2 - \theta^2)/\theta; \; \tau_1(\theta) = \theta.$$

For $\theta \in [0, 1]$:

$$I_{sup}(\theta) = \{1\}; \; x_1(\theta) = 1 - \theta - \theta^2; \; \tau_1(\theta) = \theta.$$

The variation laws of optimal feasible solution $x(\theta)$ are presented in Fig. 4.1. The variation law of the estimate α^θ and the variation law of optimal support moments $\tau_{sup}(\theta)$ are presented in Fig. 4.2.

It is seen that the a posteriori estimate coincides with the a priori one, i.e. the uncertainty of the problem cannot be decreased on this interval because of the measurement errors. After the moment $\theta = 8$ the optimal estimator is structurally stabilized ($I_{sup}(\theta) = \{2, 3\}$ for $\theta > 8$).

We note that the estimate (4.4) has the more general form, e.g.

$$\hat{\alpha}^\theta = \max_{z \in \hat{X}^\theta_*} h' x(\theta | z) .$$

The equation of the estimator (4.6) also holds for it.

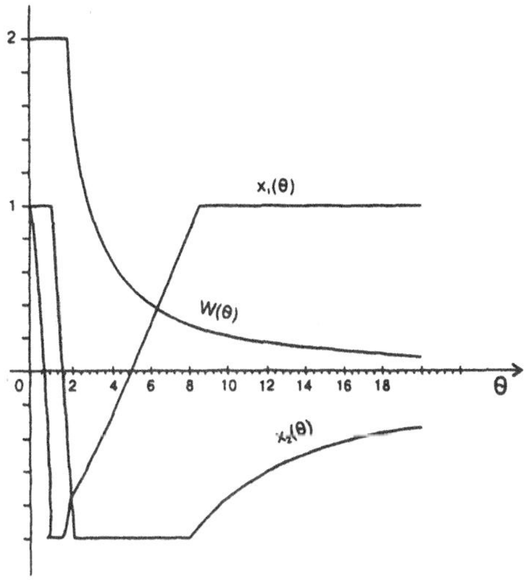

Fig. 4.1. Variation laws of the initial state and the force.

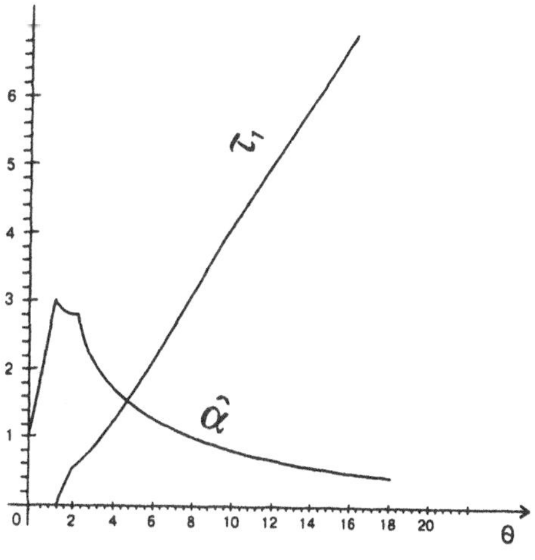

Fig. 4.2. Variation laws of the estimate and the support moment.

2.5. OPTIMAL IDENTIFICATION OF DYNAMIC SYSTEMS.

Identification of perturbations and dynamic object parameters composes an important part of the general optimal control theory [12,33].

A new approach to the identification problem of control systems is proposed in this section. It consists of receiving certain numeric characteristics of variation sets of unknown parameters by solving the extremal problem constructed in a special way.

We construct the concrete solutions to a number of identification problems on the basis of a general approach to the identification problems (see Sections 2.2 and 2.3).

2.5.1. The perturbation identification problem.

Let the dynamic system on the interval $T=[t_*,t^*]$ be described by

$$\dot{x} = A(t)x + \omega(t),\ x(t_*) = x_0 \tag{5.1}$$

where x is a n-vector of state, $A(t)$, $t \in T$, is an $n\times n$ matrix piecewise continuous function, $\omega(t)$ is a n-vector function of unknown perturbations.

Assume that a priori information on perturbatons has the form

$$\omega(t) = \omega_0(t) + \sum_{i=1}^{q} w_i\omega_i(t), \tag{5.2}$$

$$w = (w_1, w_2, \ldots, w_q) \in \overset{\vee}{W} = \{\ w \in R^q:\ Gw = f,\ w_* \le w \le w^*\ \}$$

where $\omega_0(t), \omega_1(t), \ldots, \omega_q(t), t \in T$, are known piecewise continuous functions, G is a $n\times q$ known matrix and f, w_*, w^* are known vectors.

By analogy with Section 2.4 we shall suppose that there

is the measuring device (4.2), (4.3).

Assume the measuring device (4.2) has given the signal $y(t)$, $t \in T_{\vartheta}=[t_*,\vartheta]$, where ϑ is some moment from the interval of control. This information allows us to delete from the a priori distribution $\check{W}$ of perturbation parameters the elements that were not realized automatically in the situation considered. The set $\hat{W}$ consisting of those and only those elements $w \in \check{W}$, which together with some errors $\xi(t)$, $t \in T_{\vartheta}$, are able to generate the observed signal $y(t)$, $t \in T_{\vartheta}$, will be called the a posteriori parameter distribution.

In order to apply the results of identification to linear problems of optimal control (see Sections 2.2 and 2.3) we shall consider the problem of identification of perturbations, consisting of calculation of the linear estimate

$$\hat{\alpha}_{\vartheta} = \max h'x(t^*|w), \; w \in \hat{W}, \tag{5.3}$$

where $x(t^*|w)$ is a terminal state of system (5.1), corresponding to the value w of the vector of the perturbation parameters (5.2).

Let $F(t,\tau)$, t, $\tau \in T$, be a fundamental matrix of solutions of the homogeneous part of system (5.1),

$$z(t) = y(t) - \int_{t_*}^{t} C(t)F(t,\tau)\omega_0(\tau)d\tau,$$

$$M(t) = [m_i(t), i=\overline{1,q}\;], \; \hat{h}' = (\hat{h}_i, i=\overline{1,q}),$$

$$m_i(t) = \int_{t_*}^{t} C(t)F(t,\tau)\omega_i(\tau)d\tau,$$

$$\hat{h}_i = \int_{t_*}^{t^*} h'F(t^*,\tau)\omega_i(\tau)d\tau, \; i=\overline{1,q}\;.$$

In new designations problem (5.3) takes the form:

$$\hat{\alpha}_\vartheta = \max \hat{h}'w,$$

$$\xi_* \le z(t)-M(t)w \le \xi^*, t \in T_\vartheta; \qquad (5.4)$$

$$Gw = f, \; w_* \le w \le w^*.$$

For the synthesis of optimal systems one needs to calculate the estimates (5.3) in real time. It can be done by means of an optimal identifier generating the elements of the solution to problem (5.4) continuously. In the calculations cited below let us restrict ourselves to the case $G = 0$, $f = 0$, $m = 1$.

According to [19] the optimal support feasible solution $\{w(\vartheta), S_{sup}(\vartheta)\}$, is the solution of simple problem (5.4). The support

$$S_{sup}(\vartheta) = \{\tau_{sup}(\vartheta), \; J_{sup}(\vartheta)\}$$

consists of the family $\tau_{sup}(\vartheta)$ of time moments

$$\tau_i(\vartheta), i=\overline{1,l} \; (t_* \le \tau_1(\vartheta) < \tau_2(\vartheta) < \ldots < \tau_l(\vartheta) \le \vartheta)$$

and the set of indices

$$J_{sup}(\vartheta) \subset J, \; J_{sup}(\vartheta) = l.$$

The support $l\times l$-matrix

$$M_{sup}(\vartheta) = M(\tau_{sup}(\vartheta), J_{sup}(\vartheta)) = \begin{bmatrix} M_j(\tau_i(\vartheta)), j \in J_{sup}(\vartheta), \\ i = \overline{1,l} \end{bmatrix},$$

$$\det M_{sup}(\vartheta) \ne 0,$$

the vector of potentials

$$\upsilon'(\vartheta) = \upsilon'(\tau_{sup}(\vartheta)) = \hat{h}'_{sup} M^{-1}_{sup}(\vartheta),$$

and the vector of estimates

$$\Delta'_N = \Delta'(J_N | \vartheta) = \upsilon'(\vartheta) M(\tau_{sup}(\vartheta), J_N) - \hat{h}'_N, J_N = J \backslash J_{sup}(\vartheta),$$

$$\hat{h}_{sup} = \hat{h}(J_{sup}(\vartheta)),\ \hat{h}_N = h(J_N),$$

correspond to it.

By the support S_{sup} and its accompanying vectors $\upsilon(\tau_{sup}(\vartheta))$, $\Delta(J_N\ \vartheta)$, with the help of the optimality criterion [19] we can test the feasible solution $w(\vartheta)$ for optimality.

Let ϑ be a moment of time such that:

1) the functions $A(t);C(t);\omega_i(t),i=\overline{0,q},t \in T_\vartheta$, are continuous together with the second derivatives in the neighbourhood of the points $\tau_i(\vartheta),i = \overline{1,l}$;

2) $w_{*sup} < w_{sup}(\vartheta) < w^*_{sup}$;

3) $\xi_* < z(t)-M(t)w(\vartheta) < \xi^*,\ t \in T_\vartheta\backslash\tau_{sup}(\vartheta)$;

4) $\upsilon(\tau_{sup}(\vartheta)) \neq 0,\ \Delta(J_N|\ \vartheta) \neq 0$;

5) $\tau_l(\vartheta) = \vartheta,\ \dot{z}(\vartheta)-\dot{M}(\vartheta)w(\vartheta) \neq 0$;

6) $\ddot{z}(\tau_i(\vartheta))-\ddot{M}(\tau_i(\vartheta))w(\vartheta)\neq 0,\ i=\overline{1,l-1}$.

Then the family $J_{sup}(\vartheta)$ remains unchanged in the neighbourhood of the point ϑ; the components $\tau_{sup}(\vartheta)$, $w_{sup}(\vartheta)$, of solution $\{w(\vartheta),S_{sup}(\vartheta)\}$ of problem (5.4) satisfy the system of differential equations

$$M_{sup}(\tau_i(\vartheta))\ \dot{w}_{sup} = 0, i=\overline{1,l-1};$$

$$M_{sup}(\vartheta)\ \dot{w}_{sup} = \dot{z}(\vartheta)-\dot{M}(\vartheta)w(\vartheta);\ \tau_l=1; \qquad (5.5)$$

$$(\ddot{z}(\tau_i)-\ddot{M}(\tau_i)w)\dot{\tau}_i=M_{sup}(\vartheta)\dot{w}_{sup},\ i=\overline{1,l-1}.$$

Correlations (5.5) will be called the equations of the optimal identifier of perturbations. The initial conditions for (5.5) can be obtained at $\vartheta = t_*$ using the a priori distribution of parameters $\check{W}$. The form of equations (5.5) undergoes qualitative changes in the moments of violation of conditions 1)-6).

2.5.2. The input device identification problem.

Consider the control system

$$\dot{x} = A(t)x + b(t)u \ , \ x(t_*) = x_0, \tag{5.6}$$

with an incomplete given input device $b(t)$, $t \in T$.

Assume that the function $b(t), t \in T$, has the form

$$b(t) = b_0(t) + \sum_{i=1}^{q} w_i b_i(t)$$

where $b_0(t), b_1(t), \ldots, b_q(t), t \in T$, are the known piecewise continuous functions and $w = (w_1, w_2, \ldots, w_q)$ is a q-vector from the set $\check{W}$.

The piecewise continuous control $u(t)$, $t \in T$, being known, the system (5.6) converges to (5.1). This allows us to use the results of the previous section to solve the input device identification problem.

2.5.3. The control object identification problem.

Let us investigate the system

$$\dot{x} = A(t,w)x \ , \ x(t_*) = x_0, \tag{5.7}$$

with $n{\times}n$- matrix function $A(t,w)$, $t \in T$, in the form

$$A(t) = A_0(t) + \sum_{i=1}^{q} w_i A_i(t),$$

where $A_0(t), A_1(t), \ldots, A_q(t)$, $t \in T$, are the known piecewise continuous $n{\times}n$ matrix functions, $w=(w_1, w_2, \ldots, w_q)$ is a q-vector from the set $\check{W}$.

Supply the system (5.7) with the measuring device (4.2), (4.3). The identification problem of object parameters (5.7) formally remains the same as in Section 2.4.1. But now the solution is connected with the nonlinear semi-infinite extreme problem

$$\hat{\alpha}_{\vartheta} = \max h'w,$$
$$\xi_* \le z(t) - M(t)w \le \xi^*, t \in T_{\vartheta}; \qquad (5.8)$$
$$Gw = f, \; w_* \le w \le w^*.$$

We obtain an optimal identifier for the case $G = 0$, $f = 0$, $m = 1$.

Let $w^0(\vartheta)$ be an optimal feasible solution of simple problem (5.8), $x^0(t|w^0)$, $t \in T$, be the corresponding solution of problem (5.7).

Generalizing the technique for investigation of the semi-infinite linear problems [19] and the constructive approach to solving nonlinear problems [19] we write the necessary support conditions of optimality in the problem (5.8) for $w^0(\vartheta)$.

Denote $r(w^0) = h'f(t^*, w^0)$, $M(t, w^0) = C(t)f(t, w^0)$, where the function $f(t, w^0), t \ge t_*$, is the solution to the inhomogeneous linear differential equation

$$\dot{f} = A(t, w^0)f + B(t, x^0), f(t_*, w^0) = [0]^{n\times q},$$

$$B(t, x^0) = \frac{\partial}{\partial w}(A(t, w)x^0) = [\, A_i(t)x^0, i = \overline{1, q}\,].$$

Let $S_{sup}(\vartheta) = \{\tau_{sup}(\vartheta), J_{sup}(\vartheta)\}, \tau_{sup}(\vartheta) = \{\tau_i(\vartheta) \in T_{\vartheta}, i = \overline{1, l}\,\}$,

$J_{sup}(\vartheta) \subset J$, $|J_{sup}(\vartheta)| = l$.

We shall call the totality $S_{sup}(\vartheta)$ a local support of problem (5.8) if the matrix

$$M_{sup}(\vartheta, w^0) = \left[M_j(\tau_i(\vartheta), w^0), \begin{matrix} j \in J_{sup}(\vartheta), \\ i = \overline{1, l} \end{matrix} \right]$$

is non-singular.

The vector of potentials

$$\upsilon'(\vartheta, w^0) = r'(J_{sup}(\vartheta)|w^0)M^{-1}_{sup}(\vartheta, w^0),$$

and the vector of estimates

$$\Delta'_N(\vartheta,w^0) = \upsilon'(\vartheta,w^0)M(\tau_{sup}(\vartheta),J_N|w^0) - r'_N(J_N|w^0), J_N = J\backslash J_{sup}(\vartheta),$$

are constructed by the support $S_{sup}(\vartheta)$.

Assume that ϑ is a moment of time such that for the totality $\{w^0(\vartheta), S_{sup}(\vartheta)\}$ the following conditions are fulfilled:

1) the functions $y(t); C(t); A_i(t), i=\overline{0,q}$, $t \in T_\vartheta$, are continuous together with second derivatives in the neighborhood of the points $\tau_i(\vartheta), i = \overline{1,l}$;

2) $w_{*sup} < w^0_{sup}(\vartheta) < w^*_{sup}$;

3) $\xi_* < y(t)-C(t)x^0(t|w^0) < \xi^*$, $t \in T_\vartheta \backslash \tau_{sup}(\vartheta)$.

Then for optimality of the feasible solution $w^0(\vartheta)$ in problem (5.10) the fulfilment of correlations

$$\Delta_j(\vartheta,w^0) \le 0 \quad \text{for} \quad w^0_j = w_{*j} \; ; \; \Delta_j(\vartheta,w^0) \ge 0 \quad \text{for} \; w^0_j = w^*_j \; , j \in J_N$$

$$\Delta_j(\vartheta,w^0) = 0 \quad \text{for} \quad w_{*j} < w^0_j(\vartheta) < w^*_j;$$

$$\upsilon_i(\vartheta,w^0) \ge 0 \text{ for } y(\tau_i)-C(\tau_i)x^0(\tau_i|w^0)=\xi^* ;$$

$$\upsilon_i(\vartheta,w^0) \le 0 \text{ for } y(\tau_i)-C(\tau_i)x^0(\tau_i|w^0)=\xi_* ;$$

$$\upsilon_i(\vartheta,w^0)=0 \text{ for } \xi_*<y(\tau_i)-C(\tau_i)x^0(\tau_i|w^0)<\xi^*, i =\overline{1,l}.$$

is necessary.

Supply the correlations 1)-3) with conditions

4) $\upsilon(\vartheta,w^0)\ne 0$, $\Delta_N(\vartheta,w^0)\ne 0$;

5) $\tau_l(\vartheta)=\vartheta$, $\dot{y}(\vartheta) - (\dot{C}(\vartheta) + C(\vartheta)A(\vartheta,w^0))x^0(\vartheta|w^0)\ne 0$;

6) $\ddot{y}(\tau_i) - [\ddot{C}(\tau_i)+2\dot{C}(\tau_i)A(\tau_i,w^0)+C(\tau_i)(\dot{A}(\tau_i,w^0)+A^2(\tau_i,w^0))]\times$

$\times x^0(\tau_i|w^0)\ne 0$, $i=\overline{1,l-1}$.

Conditions 1)-6) in the neighbourhood ϑ being fulfilled, equations of optimal identifier have the form

$$\dot{x} = A(t,w)x \ ; \ \dot{f} = A(t,w)f + B(t,x);$$

$$M_{sup}(\tau_i,w)\ \dot{w}_{sup} = 0, i=\overline{1,l-1};$$

$$M_{sup}(\vartheta,w)\dot{w}_{sup}=y(\vartheta)-(\dot{C}(\vartheta)+C(\vartheta)A(\vartheta,w))x(\vartheta|w);$$

$$[\ddot{y}(\tau_i)-[\ddot{C}(\tau_i)+2\dot{C}(\tau_i)A(\tau_i,w)+C(\tau_i)(\dot{A}(\tau_i,w)++A^2(\tau_i,w))]\times$$

$$\times x(\tau_i|w)]\dot{\tau}_i=\dot{M}_{sup}(\tau_i,w)\dot{w}_{sup};$$

$$\tau_l=1.$$

Example 5.1. Let us illustrate the results for the perturbation identification problem

$$\omega(t) = w_1(t)t+w_2(t)$$

of the dynamic system

$$\dot{x}_1 = x_2, \ \dot{x}_2 = \omega(t), \ x_1(0) = 0, \ x_2(0) = 0,$$

$$0 \le w_1 \le 1, \ \ -1 \le w_2 \le 0.$$

Use the measuring device $y = x_2 + \xi, 0 \le \xi \le 1$. Assume that it registers the signal $y(t)\equiv 4$. Calculate the estimate

$$\hat{\alpha}_\vartheta = \max_{w\in\hat{W}} h'x(6|w).$$

For $h = (1/2,-2/9)$ the problem has the form

$$\max w_1-w_2, \ -1 \le w_1t^2/2 + w_2t \le 0, \ t\in[0,\vartheta],$$

$$0 \le w_1 \le 1, \ -1 \le w_2 \le 0.$$

Conditions 1)-6) are fulfilled for the moment $\vartheta=2$. The optimal identifier for $2 \le \vartheta \le 4$ has the form

$$w_1\vartheta^2/2 + w_2\vartheta = \vartheta, \ \tau=\vartheta, \ w_1(2)=1, w_2(2)=-1.$$

Hence, $w_2^o(\vartheta) = -1$, $w_1^o(\vartheta) = 2/\vartheta$. Condition 2) is violated at the point $\vartheta = 4$. For $\vartheta \geq 4$ the optimal identifier is described by

$$\dot{w}_1\tau^2/2 + \dot{w}_2\tau = 0,\ w_1\vartheta^2/2 + w_2\vartheta = \vartheta,$$

$$-w_1\dot{\tau} = \tau\dot{w}_1 + \dot{w}_2, w_1(4)=1/2, w_2(4)=-1.$$

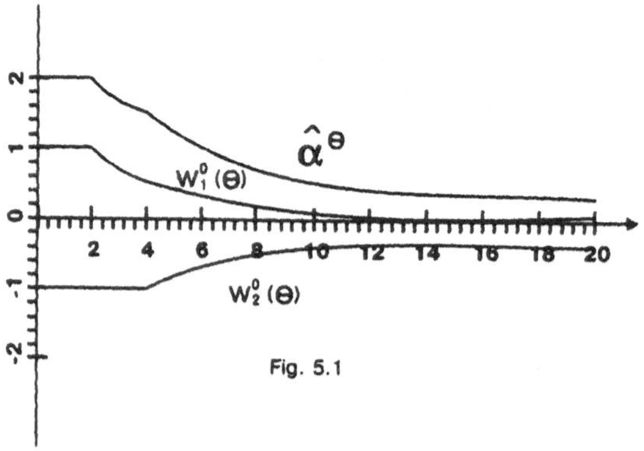

Fig. 5.1

The solution to this system has the form $\tau = \vartheta/2$, $w_1^o(\vartheta) = 8/\vartheta^2, w_2^o(\vartheta) = -4/\vartheta$. The variation laws of the parameters $w_1^o(\vartheta)$, $w_2^o(\vartheta)$, and the estimate $\hat{\alpha}_\vartheta$ are given in Fig. 5.1.

CHAPTER 3

OPTIMAL CONTROLLERS

3.1. OPTIMAL CONTROLLER FOR DISCRETE CONTROL SYSTEMS.

The problem of synthesis of optimal systems is a central one in optimal control theory. Different methods have been used for its solution [4,7,9,14,36,37]. However, until now satisfactory results have not been obtained for multidimensional problems with direct restrictions on control.

We are going to solve the problem of synthesis for discrete control systems. In this case one can use also the method of dynamic programming [4]. But in practice such a solution cannot be realized with high accuracy if the systems are of third order and higher because of the necessity for a large volume of computer storage ("curse of dimension"). The method of construction of switching surfaces using the maximum principle also requires a large volume of computer storage of for recording the relations describing these surfaces [36,37]. The method presented in this chapter is based on adaptation of the dual methods of linear programming (LP) to optimization of dynamic systems [15-22].

Contrary to two mentioned "static" synthesis methods the proposed approach is "dynamic". Both in dynamic programming and in the method of switching surfaces the optimal feedback control $u(x)$ and the switching surfaces equations $S(x)=0$ are constructed before the system starts operating. In the real control process only the current system states are measured. It is thus necessary to carry out an enormous amount of preliminary work to compute the optimal controls for all system states, much of which will not occur in the control process. But the main defect is not so much in the amount of preliminary work, as in the necessity for storage of the results of all this work. In the "dynamic" method of optimal systems synthesis given below, the construction of the optimal control is distributed over the whole period of the control process. At each particular moment our controller constructs an optimal control for the current state where the system transfers from the previous one under the influence of the control and perturbation realized at this moment. This allows us to save enormous amount of memory at the expense of relatively little computational work in the control process. The advantage in the amount of current calculations is achieved because the optimal control is not constructed anew for each current state, but results from correction of the old optimal control in response to continuous perturbations affecting the system trajectory.

It is known from linear programming that the dual method is an extremely effective one for correction of optimal feasible solutions by slight changes of problem parameters. That is why the proposed approach is based on special implementation of the dual method for solving optimal control problems.

At the same time, small variations in the parameters are the consequence of constantly (with the small quantization period $h > 0$) measuring the current system state. In this section a classical case of synthesis is considered assuming that the exact measurements of system states are available. It is made in order to explain better the main idea of the proposed approach. Controllers using incomplete and inexact state measurements are constructed in Chapter 4.

The proposed algorithm for the controller work is intended for utilization of microprocessor devices. The description of this algorithm (Section 3.1.2) gives the memory volume used and the number of arithmetic operations at each step without increasing computer expense, with the rising demands for accuracy in obtaining the result.

3.1.1. Statement of the problem.

We consider a discrete system, the behaviour of which on $T(t_*) = \{ t_*, t_* + h, t^* - h \}$ is described by

$$x(t) = A(t,h)x(t) + b(t,h)u(t),\ x(t_*) = x_0 \qquad (1.1)$$

where $x(t)$ is the n-vector state of system (1.1) at the moment t and $u(t)$ is the value of a scalar control.

We shall assume that the $n \times n$-matrix $A(t,h), t \in T(t_*)$, and the n-vector $b(t,h)$, $t \in T(t_*)$, were obtained either by discretization of the continuous system

$$\dot{x} = A(t)x + b(t)u$$

or after transition to impulse controls. In the first case the simplest Euler method leads to

$$A(t,h) = E + hA(t),\ b(t,h)=hb(t).$$

The second one gives

$$A(t,h) = F(t+h,t),$$

$$b(t,h) = \int_t^{t+h} F(t+h,\tau)b(\tau)d\tau,$$

$$(\partial F(t,\tau)/\partial t=A(t)F(t,\tau),\ F(\tau,\tau)=E).$$

In view of what was said above, the assumption

$$\det A(t,h) = 0, t \in T(t_*),$$

is not an essential restriction.

As usual the sequence $u(t)$, $t \in T(t_*)$, restricted by

$$u_*(t) \le u(t) \le u^*(t),\ t \in T(t_*), \qquad (1.2)$$

will be called a (program) control.

Let in the state space the terminal set

$$X^* = \{x \in R^n : Hx = g\}, g \in R^m,\ \mathrm{rank}\ H = m < n \qquad (1.3)$$

be given.

A control $u(t)$, $t \in T(t_*)$, will be called admissible for position $\{t_*, x(t_*)\}$ if the corresponding trajectory $x(t)$, $t \geq t_*$, of system (1.1) at t^* reaches the set

$$x(t^*) \in X^* . \quad (1.4)$$

The admissible control $u(t)$, $t \in T(t_*)$, is said to be an optimal one $(u^0(t), t \in T(t_*))$, if the quality criterion

$$J(u) = c'x(t^*) \quad (1.5)$$

attains the maximal value.

A finite algorithm for constructing the optimal control $u^0(t|t_*, x_0), t \in T(t_*)$, for a fixed initial position $\{t_*, x_0\}$ is presented in a monograph [19]. If perturbations do not affect system (1.1) then the constructed program control solves the synthesis problem completely, since for an arbitrary moment $\tau \in T(t_*)$ system (1.1) is in state $x^0(\tau)$ and the optimal control value for position $\{\tau, x^0(\tau)\}$ is equal to $u^0(\tau|t_*, x_0\}$. The synthesis problem assumes implicitly that the discrete system is subjected to the action of perturbations. If the latter are known the synthesis problem is not complicated in principle.

Unknown perturbations may be of differing natures:

1) random perturbations with known probability characteristics which do not depend on the choice of control $u(t)$, $t \in T(t_*)$;

2) indefinite ones with known sets of values which do not depend on the control $u(t)$, $t \in T(t_*)$;

3) playing perturbations that depend on the choice $u(t)$, $t \in T(t_*)$.

We consider a type of perturbation which is linked with the classical statement of the problem of synthesis directly. We shall consider that as a result of control actions

$$u(t), \ t_* \leq \tau \leq \tau - h$$

and perturbations

$$y(t,h),\ t_* \le \tau \le \tau-h$$

the system (1.1), being at $t = \tau$ in the state $x(\tau)$, is found at $t = \tau + h$ with the perturbation $y(\tau,h)$, not in the state

$$\check{x}(\tau+h)= A(\tau,h)x(\tau) + b(\tau,h)u(\tau)$$

but in

$$x(\tau+h)=\check{x}(\tau+h)+y(\tau+h).$$

Construction of the optimal control $u^o(\tau+h|\tau+h,x(\tau+h))$ for the position $\{\tau+h,x(\tau+h)\}$ for any $x(\tau+h)$ at any moment $\tau+h$ will be called the synthesis of the optimal discrete system (construction of optimal controller).

This statement of the problem of synthesis relies on the program solution to problem (1.1)-(1.5) and requires an optimal parrying perturbation acting on the discrete system trajectory emanating from the state x_o at the moment t_*.

3.1.2. Optimal controller.

Let $\tau \in T(t_*)$ is an arbitrary moment in time; $x(\tau)$ is the state of the system (1.1) at this moment. Denote by

$$\{\ u^o(t|\tau,x(\tau)),\ t \in T(\tau),\ t \in T(\tau);\ T_{sup}(\tau)\}$$

the solution to problem (1.1)-(1.5) where the set $T(t_*)$ and the initial state x_o are replaced by $T(\tau)$ and $x(\tau)$.

A new element $T_{sup}(\tau)$ is called an optimal support (see Chapter 1). The support $T_{sup}(\tau)$ consists of m support moments

$$\tau \le \tau_1(\tau) < \tau_2(\tau) < \dots < \tau_m(\tau) < t^*-h$$

that possess the following property. Let

$$F_h(t,\tau)A(\tau,h),\ t,\tau\in T(t_*),$$

be a fundamental matrix of solutions to the homogeneous part of discrete system

$$F_h(t,\tau-h) = F_h(t,\tau)A(\tau,h),\ F_h(t,t-h) = E;$$
$$F_h(t+h,\tau) = A(t,h)F_h(t,\tau), F_h(\tau+h,\tau) = E\ ).$$

Compose the $m\times m$-matrix

$$P(\tau) = (\ HF_h(t^*,t)b(t,h),\ t \in T_{sup}(\tau)\ ). \qquad (1.6)$$

The totality $T_{sup}(\tau)$ is called a support if $det\ P(\tau) = 0$. Construct the vector of potentials

$$\upsilon'(\tau) = c'_{sup}Q(\tau) \qquad (1.7)$$

where

$$c_{sup} = (\ c(t),\ t \in T_{sup}(\tau)\ ),\ c(t) = c'F_h(t^*,t)b(t,h),\ t \in T(t_*),$$
$$Q(\tau) = P^{-1}(\tau)$$

corresponding to the support $T_{sup}(\tau)$.

The co-trajectory $\psi(t) = \psi(t|\tau),\ t \in T(\tau)$, that accompanies the support $T_{sup}(\tau)$ will be called the solution to the conjugate system

$$\psi'(t-h) = \psi'(t)A(t,h),\ \psi'(t^*-h) = c' - \upsilon'(\tau)H.$$

It generates the co-control

$$\Delta(t) = \Delta(t|\tau),\ t \in T(\tau)\ :\ \Delta(t) = -\psi'(t)b(t,h),\ t\in T(\tau). \qquad (1.8)$$

With respect to construction at the support moments the co-control (1.8) equals zero:

$$\Delta(t)=0, t\in T_{sup}(\tau)$$

The optimal control at the non-support moments $t\in T_N(\tau) = T(\tau)\setminus T_{sup}(\tau)$ has the form

$$u^0(t|\tau,x(\tau)) \begin{cases} =u_*(t) \text{ when } \Delta(t)>0; \\ =u^*(t) \text{ when } \Delta(t)<0; \\ \in[u_*(t),u^*(t)] \text{ when } \Delta(t)=0, t\in T_N(\tau). \end{cases} \tag{1.9}$$

The totality of values of the optimal control at the support moments $u^0_{sup}=(u^0(t|\tau,x(\tau)),t\in T)$ is calculated by formula

$$u^0_{sup}=Q(\tau)g(\tau), \tag{1.10}$$

where

$$g(\tau) = g - HF_h(t^*,t_* - h)x_0 - \sum_{t=t_*}^{\tau-h} HF_h(t^*,t)y(t,h) -$$

$$- \sum_{t=t_*}^{\tau-h} HF_h(t^*,t)b(t,h)u^0(t|t,x(t)) -$$

$$- \sum_{t\in T_N(\tau)} HF_h(t^*,t)b(t,h)u^0(t|\tau,x(\tau))$$

In the future we will need the following additional information about the optimal support control

Construct the sets

$$T_{N+}(\tau) = \{ t \in T_N(\tau): \Delta(t) > 0 , \Delta(t)\Delta(t-h) < 0\} \quad \cup$$

$$\cup \{ t \in T_N(\tau): \Delta(t) > 0, t - h \in T_{sup}(\tau)\},$$

$$T_{N-}(\tau) = \{ t \in T_N(\tau): \Delta(t) < 0, \Delta(t)\Delta(t-h) < 0 \} \quad \cup$$

$$\cup \{ t \in T_N(\tau): \Delta(t) < 0, t - h \in T_{sup}(\tau)\}.$$

We limit ourselves to the case where the number of elements $|T_{N+}(\tau)|+|T_{N-}(\tau)|$ exceeds $T_{sup}(\tau)$ by not more than unity, i.e. it equals $m+1$.

Assume that we know the information array

$$F_h(t^*,t), \ t \in T_{N+}(\tau) \cup T_{N-}(\tau) \cup \tau \cup (t^*-h).$$

Let us pass to the description of the algorithm for the optimal controller for position $\{\tau+h,x(\tau+h)\}$. The operations given below are the implementation of the dual method of LP problem solution for the situation arising in optimal control problems (see Chapter 1 and the Appendix).

The set

$$C^k(\tau+h) = \{ u^{(k)}(t), \ t \in T(\tau+h) \ ; \ T^k_{sup} = \{ \tau_1,\dots,\tau_m \};$$

$$T^k_{N+}; \ T^k_{N-}; \ y^k; \ \Phi^k(t), \ t \in T^k_{N+} \cup T^k_{N-} \cup \tau \cup (t^* - h);$$

$$\psi^k(t), \ t \in T^k_{N+} \cup T^k_{N-} \cup (t^* - h); \ Q^k \}$$

will be called the current state of the algorithm at the moment $\tau+h$. As an initial state $C^0(\tau+h)$ we shall choose the set with the following components:

$$u^{(0)}(t) = u(t|\tau,x(\tau)), \ t\in T(\tau+h); \ T^0_{sup} = T_{sup}(\tau) = \{\tau_1,\dots,\tau_m\};$$

$$T^0_{N+}=T_{N+}(\tau); \quad T^0_{N-}=T_{N-}(\tau); \quad y^0=x(\tau+h)-A(\tau,h)x(\tau)-b(\tau,h)u(\tau);$$

$$\Phi^0(t) = F_h(t^*,t), \ t \in T_{N+}(\tau) \cup T_{N-}(\tau) \cup \tau \cup (t^* - h);$$

$$\psi^0(t) = \psi(t), \ t \in T_{N+}(\tau) \cup T_{N-}(\tau) \cup (t^* - h); \ Q^0=Q(\tau).$$

An iteration of the algorithm $C^k(\tau+h) \longrightarrow C^{k+1}(\tau+h)$ $(C^k(\tau+h) \longrightarrow C^0(\tau+2h))$ consists of the following steps

Step 1. Compare τ with τ_1. If $\tau < \tau_1$, then pass to *Step 2*, for $\tau = \tau_1$ we pass to *Step 9*.

Step 2. Calculate the vector

$$\Delta u^k(T^k_{sup}) = (\Delta u^k(t),\ t \in T^k_{sup}) = -Q^k H \Phi^k(\tau) y^k,$$

$$\Delta u^k(T^k_N) = 0,\ T^k_N = T(\tau+h) \setminus T^k_{sup}.$$

Step 3. Calculate the number μ^k and find the moment $\tau_s \in T^k_{sup}$:

$$\mu^k = \mu(\tau_s) = \min \mu(t),\ t \in T^k_{sup};$$

$$\mu(t) = \begin{cases} -\dfrac{u_*(t) - u^{(k)}(t)}{\Delta u^k(t)}, & \text{when } \Delta u^k(t) < 0; \\ \dfrac{u^*(t) - u^{(k)}(t)}{\Delta u^k(t)}, & \text{when } \Delta u^k(t) > 0; \\ \infty, & \text{when } \Delta u^k(t) = 0, t \in T^k_{sup}. \end{cases}$$

For $\mu^k \geq 0$ we pass to *Step 4*. If $\mu^k < 1$ then we pass to *Step 5*.

Step 4. Assume $u^0(\tau+h|\tau+h,\ x(\tau+h)) = u^{(k)}(\tau+h) + \Delta u^k(\tau+h)$ If $\tau + h = t^* - mh$ then the algorithm completes the work:

$$u^0(\tau+ih|\tau+ih, x(\tau+ih)) = u^{(k)}(\tau+ih) + \Delta u^k(\tau+ih),\ i=\overline{1,m}.$$

If $\tau + h < t^* - mh$ we construct an initial state $C^0(\tau + 2h)$ for the moment $\tau + 2h$ with the following components:

$$u^{(0)}(t) = u^{(k)}(t) + \Delta u^k(t),\ t \in T(\tau+2h);\ T^0_{sup} = T^k_{sup};$$

$$T^0_{N+} = T^k_{N+};\ T^0_{N-} = T^k_{N-};\ y^0 = x(\tau+2h) - A(\tau+h,h)x(\tau+h) -$$

$$- b(\tau+h,h)u(\tau+h);$$

$$\Phi^{0}(t) = \Phi^{k}(t),\ t \in T^{0}_{N+} \cup T^{0}_{N-} \cup (t^{*} - h);$$

$$\Phi^{0}(\tau + h) = \Phi^{k}(\tau)A^{-1}(\tau+h,h);$$

$$\psi^{0}(t) = \psi^{k}(t),\ t \in T^{0}_{N+} \cup T^{0}_{N-} \cup (t^{*} - h);\ Q^{0}=Q^{k}\ .$$

Pass to *Step* 1.

Step 5. Calculate

$$\Delta^{k}(t) = -\psi^{k}(t)'b(t,h),$$

$$\xi^{k}(t)' = \rho q^{k}(\tau_{s})H\Phi^{k}(t),$$

$$\delta^{k}(t) = \xi^{k}(t)'b(t,h),\ t \in T^{k}_{N+} \cup T^{k}_{N-} \cup (t^{*}-h),$$

where $q^{k}(\tau_{s})$ is the line of the matrix Q^{k} corresponding to the moment τ_{s}, $\rho = -sign\ \Delta u^{k}(\tau_{s})$. Pass to *Step 6*.

Step 6. Calculate the numbers

$$\sigma^{k} = \sigma(t_{q}) = min\ \sigma(t),\ t \in T^{k}_{N+} \cup T^{k}_{N-} \cup (t^{*}-h);$$

$$s(t),\ t \in T^{k}_{N+} \cup T^{k}_{N-}:$$

$$\sigma(t) = -\frac{\Delta^{k}(t)}{\delta^{k}(t)},\ s(t) = 0,\ \text{when}\ \ t \in T^{k}_{N+},\ \delta^{k}(t) < 0\ \text{or}\ t \in T^{k}_{N-},$$

$$\delta^{k}(t)>0;\ \sigma(t) = -\frac{\Delta^{k}(t-h)}{\delta^{k}(t-h)} = -\frac{\psi^{k}(t)'A(t,h)b(t-h,h)}{\xi^{k}(t)'A(t,h)b(t-h,h)},\ s(t) = h,$$

when $t \in T^{k}_{N+}$, $\delta^{k}(t-h) < 0$ or $t \in T^{k}_{N-}$, $\delta^{k}(t - h) > 0$,

$$t - h \in T^{k}_{sup};$$

$$\sigma(t) = -\frac{\Delta^{k}(t-2h)}{\delta^{k}(t-2h)} = -\frac{\psi^{k}(t)'A(t,h)A(t-h,h)b(t-2h,h)}{\xi^{k}(t)'A(t,h)A(t-h,h)b(t-2h,h)},\quad s(t) = 2h,$$

when $t \in T^{k}_{N+}$, $\delta^{k}(t-2h)<0$ or $t \in T^{k}_{N-}$, $\delta^{k}(t-2h) > 0$,

$t-2h \in T^{k}_{sup}$, $t-h \in T^{k}_{sup}$; $\sigma(t^{*}-h) = -\dfrac{\Delta^{k}(t^{*}-h)}{\delta^{k}(t^{*}-h)}$, when $\Delta^{k}(t^{*}-h)\delta^{k}(t^{*}-h) < 0$; $\sigma(t) = \infty$ in other cases. For $\sigma^{k}=\infty$ pass to *Step* 8. Let $\sigma^{k} < \infty$. Then pass to the next step.

Step 7. In the support T^{k}_{sup} we replace the moment τ_{s} by $t_{q} - s(t_{q})$: $T^{k+1}_{sup} = (T^{k}_{sup} \setminus \tau_{s}) \cup (t_{q} - s(t_{q}))$. We renew the matrix Q^{k} with respect to the recurrent formulae:

$$Q^{k+1}(\tau_{i},j) = Q^{k}(\tau_{i},j) - Q^{k}(\tau_{s},j)\ r^{k}(\tau_{i}) \ /\ r^{k}(\tau_{s}), i = s;$$

$$Q^{k+1}(\tau_{s},j) = Q^{k}(\tau_{s},j)/r^{k}(\tau_{s}),\ i = \overline{1,m},\ j = \overline{1,m},$$

where $r^{k} = (r^{k}(\tau_{i}),\ i = \overline{1,m}) = Q^{k}H\Phi^{k}(t_{q}-s(t_{q}))b(t_{q}-s(t_{q}))$;

$$\Phi^{k}(t_{q}-s(t_{q})) = \Phi^{k}(t_{q}) \text{ when } s(t_{q}) = 0;$$

$$\Phi^{k}(t_{q} - s(t_{q})) = \Phi^{k}(t_{q})A(t_{q},h) \text{ when } s(t_{q}) = h;$$

$$\Phi^{k}(t_{q}-s(t_{q})) = \Phi^{k}(t_{q})A(t_{q},h)A(t_{q}-h,h) \text{ when } s(t_{q})=2h.$$

Assume

$$u^{(k+1)}(t) = u^{(k)}(t) + \mu^{k}\Delta u^{k}(t),\ t \in T(\tau+h);\ y^{k+1} = (1-\mu^{k})y^{k};$$

$$\psi^{k+1}(t) = \psi^{k}(t) + \sigma^{k}\xi^{k}(t),\ t \in T^{k}_{N+} \cup T^{k}_{N-} \cup (t^{*}-h);$$

$$\psi^{k+1}(t_{q}-s(t_{q})+h)' = \psi^{k+1}(t_{q}-s(t_{q}))'A^{-1}(t_{q}-s(t_{q})+h,h);$$

$\psi^{k}(t_{q}-s(t_{q}))=\psi^{k}(t_{q})$ when $s(t_{q})=0$; $\psi^{k}(t_{q}-s(t_{q}))=\psi^{k}(t_{q})A(t_{q},h)$

when $s(t_{q}) = h$; $\psi^{k}(t_{q}-s(t_{q})) = \psi^{k}(t_{q})A(t_{q},h)A(t_{q}-h,h)$ when

$s(t_q) = 2h;\quad \Delta^{k+1}(t_q - s(t_q)+h) = -\psi^{k+1}(t_q - s(t_q)+h)'b(t_q - s(t_q)+h),$

moment (τ_s+h) exclude from $T^k_{N+} \cup T^k_{N-}$ and add correspondingly the moment $t_q - s(t_q) + h$ in T^{k+1}_{N+} if $\Delta^{k+1}(t_q - s(t_q)+h)>0$ or in T^{k+1}_{N-} if $\Delta^{k+1}(t_q - s(t_q)+h)<0$:

$$T^{k+1}_{N+} \cup T^{k+1}_{N-} = (T^k_{N+} \cup T^k_{N-}) \setminus (\tau_s+h) \cup (t_q - s(t_q) + h),$$

$$(t_q - s(t_q) + h) \in T(\tau+h);$$

$$\Phi^{k+1}(t) = \Phi^k(t)\ ,\ t \in (T^{k+1}_{N+} \cup T^{k+1}_{N-}) \setminus (t_q - s(t_q)+h),$$

$$\Phi^k(t_q - s(t_q) + h) = \Phi^k(t_q - s(t_q))A^{-1}(t_q - s(t_q) + h),h).$$

Pass to *Step 2*.

Step 8. The situation $\sigma^k = \infty$ testifies that the arising perturbation $y(\tau,h)$ does not allow us to satisfy the terminal constraints $Hx(t^*) = g$ at the expense of choice of control $u(t)$ on the segment $T(\tau+h)$ at the moment t^*. The algorithm stops.

Step 9. Assume $s = 1$, $\mu^0 = 0$, $\Delta u^0(\tau_1) = -q^0(\tau_1)H\Phi^0(\tau)y^0$, and pass to *Step 5*.

3.1.3. Generalizations.

The problem of synthesis in a general case cannot be solved for optimal control problems in discrete systems on a fixed interval $T(t_*)$ for all moments $\tau \in T(t_*)$. For $\tau > t^* - mh$ the discrete system becomes uncontrollable and is not able to parry perturbations which are arbitrary with respect to direction and arbitrarily small with respect to values. Hence the constructed controller relies on sufficiently small perturbations $y(t,h), t \in T(t_*)$, possessing the property

$$y(t,h) = 0,\ t_* \le t \le t^* - mh,\ y(t,h) = 0,\ t \ge t^* - mh.$$

Because of the boundedness of the control actions the operation of the controller can be interpreted before the moment t^*-mh. An increase in efficiency of the controller can be achieved either by choosing in an optimal way the moment $t^*(\tau)$ of finishing process or by making t^* mobile, $t^*=\tau+\vartheta$, where ϑ is a sufficiently large number. The synthesis algorithm is easy to generalize with respect to these cases. In connection with mobility of the terminal moment we may consider that elements c,g,H of the problem depend on t^* as well.

The method for constructing the optimal controller described above presupposes certain properties of co-control. Cases where the sets $T_{N+}(\tau)\cup T_{N-}(\tau)$ contain essentially more elements than the set $T_{sup}(\tau)$, and where in the process of operation of the controller there appear new elements in it which differ from those described above, can be investigated according to the mentioned scheme, but now the controller becomes more complicated.

Example 1.1. As an illustration of the results we consider the problem

$$x_2(3) \longrightarrow max,\ x_1(t+h) = x_1(t) + hx_2(t),\ x_2(t+h) = x_2(t)+hu(t),$$

$$x_1(3) = 1/5,\ x_1(0) = x_2(0) = 0,\ 0 \le u(t) \le 1, \qquad (1.11)$$

$$t \in \{ 0,\ 0.5,\ \ldots,\ 2.5\} ,\ h = 0.5.$$

This model is formed from the continuous problem of optimal control of the movement of the material point

$$x(3) \longrightarrow max,\ \dot{x} = u,\ x(3) = 1/5,\ x(0) = \dot{x}(0) = 0,$$

$$0 \le u(t) \le 1, t\in[0,3],$$

if the derivative $\dot{x}(t)$ is substituted in the Euler rule by the correlation

$$(x(t+h)-x(t))/h.$$

The problem (1.11) is a particular case of the problem (1.1)-(1.5)

$$A(t,h)=A(h)=\begin{bmatrix} 1 & h \\ 0 & 1 \end{bmatrix}, b(t,h)=b(h)=\begin{bmatrix} 0 \\ h \end{bmatrix}, t_*=0, t^*=3,$$

$$t_*=0, t^*=3, u_*(t)=0, u^*(t)=1, H=[1,0], c=[0,1].$$

The optimal control $u^0(\cdot|0,0)=(0,0,0,0,4/5)$ and the trajectory in the problem (1.11) are presented in Fig. 1.1.

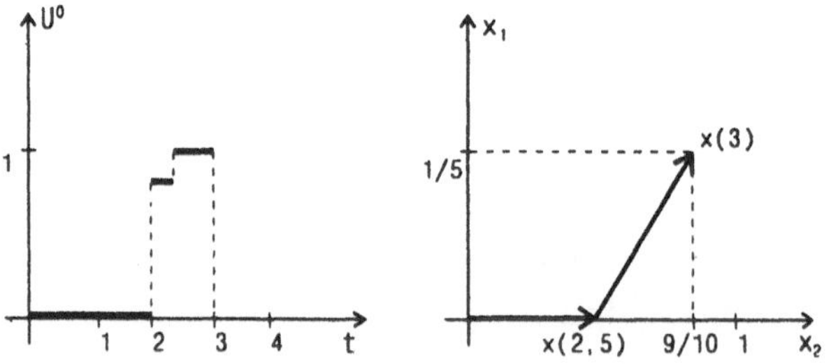

Fig.1.1

The optimal support consists of one moment $\tau_1=2$. Construct the optimal controller. The initial state at the moment $\tau=0.5$ has the form

$$C^0(0.5) = \{u^{(0)}(t) = u^0(t|0,0),\ t \in [0,5,2.5]\ ;\ T^0_{sup} = 2,$$

$$T^0_{N-} = 2.5;\ T^0_{N+} = 0;$$

$$\Phi^0(0)=\begin{bmatrix} 1 & 3 \\ 0 & 1 \end{bmatrix}, \Phi^0(2.5)=\begin{bmatrix} 1 & 0.5 \\ 0 & 1 \end{bmatrix}, y^0=-1/2; \psi^0(2.5)=\begin{bmatrix} -2 \\ 1 \end{bmatrix}, Q^0=4\}.$$

As a result of operation of the controller in the mode of real time under perturbations

$$y(0) = (1/2,0),\ y(0.5) = (-1/2,0),\ y(1)=(1/4,0),$$

$$y(1.5) = y(2) = (0,0)$$

we shall have the control $u_s^o(\cdot)$ presented in Fig. 1.2.

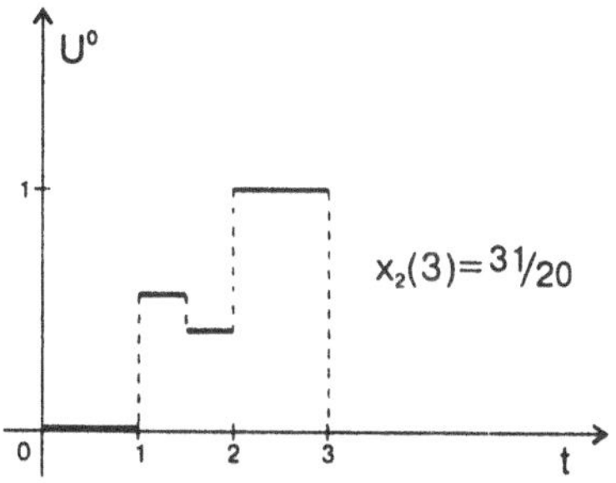

Fig.1.2

If we had access to information on perturbations before the beginning of control process, the optimal control $u_p^o(\cdot)$ would have the form represented in Fig. 1.3.

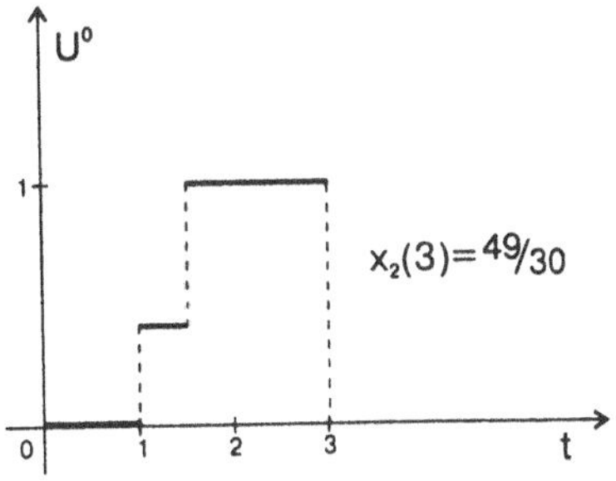

Fig.1.3

We can see from Figs 1.2 and 1.3 that the efficiency of

optimal control increases by 1/12 if the information about perturbation is known over the whole control period.

In conclusion we would like to pay attention to the following. The value of the control is noncritical in the program optimal controls of Figs. 1.1, 1.3 only at one moment. The control produced by the controller in Fig. 1.2 has the noncritical value at two moments of time.

3.2. TIME OPTIMAL SYNTHESIS OF DISCRETE CONTROL SYSTEMS.

Optimization according to a time optimal criterion is natural for the very wide class of control objects. The mathematical model of this problem contains a minimal number of additional elements (functions). It is probably not accidental that in engineering statements of the problem corresponding to the traditions of automatic regulation it was implied that the question was about closed loop optimal systems and optimal feedback control.

In this section the synthesis problem is solved for the time optimal problem in discrete systems obtained from continuous dynamic systems with a real spectrum. We use a special approximation when the actual behaviour of two systems do essentially not differ.

3.2.1. The time optimal control problem.

Consider the discrete system

$$x(t+h) = A(h)x(t) + b(h)u(t),\ x(0)=x_0. \tag{2.1}$$

Concerning the $n \times n$-matrix $A(h)$ and the n-vector $b(h)$ we suppose that they were obtained either as a result of discretization of the continuous system

$$x = Ax + bu$$

or using impulse functions $u(t) = u_k,\ t \in [\ kh,\ (k+1)h\ [$, $k = 0,1,2,\ldots$. In the first case the simplest Euler method leads to

$$A(h) = E + hA,\ b(h) = hb,$$

the second one gives

$$A(h) = \exp Ah,\ b(h) = \int_0^h e^{A(h-\tau)} b d\tau.$$

We called the sequence $u(0),\ u(h),\ u(2h),\ldots$ an admissible control for the state x_0 if it satisfies the constraints

$$|\ u(t)\ | \le 1,\ t \ge 0, \tag{2.2}$$

and generates the trajectory $x(h),\ x(2h),\ \ldots,$ of system (2.1) which at the final moment $t^* = t^*(x_0)$ becomes zero

$$x(t^*) = 0. \tag{2.3}$$

The time optimal control problem in the program statement consists of in constructing the optimal control $u^o(t) = u^o(t\,|\,t=0, x_0),\ t \ge 0$, guaranteeing (2.3) in the shortest time t^{*o}.

A synthesis optimal system usually implies the construction of function $u(x),\ x \in R^n$, satisfying the inequality

$$|u(x)| \le 1,\ x \in R^n,$$

which after substitution in (2.1) becomes the system

$$x(t+h) = A(h)x(t) + b(h)u(x(t))$$

with trajectories having the property (2.3) with the minimal possible $t^* = t^{*o}(x)$ for every initial state $x(0) = x$ from the set of controllability of system (2.1).

The given classical statement of the synthesis problem is in a certain sense static. There is no hint concerning the action of perturbations upon system (2.1), owing to which the problem of feedback control appears.

Let the control process for system (2.1) start at $t=0$

from the state x_0 and at $t=\tau$ let the discrete system be at $x(\tau)$. If at that moment we put the control $u(\tau)$ to the system then at the next moment $\tau+h$ it will be not in the state

$$\check{x}(\tau+h)= A(h)x(\tau) + b(\tau,h)u(\tau)$$

because of the perturbations $y(\tau,h)$ which appeared at τ , but in the state

$$x(\tau+h)=\check{x}(\tau+h)+y(\tau+h).$$

Statement of the problem. The problem of calculation at any $y(t,h) \in R^n$ of the optimal time $t^{*o}(x(\tau + h))$ and the value $u^o(\tau+h|\tau+h,x(\tau+h))$ of the optimal control for a new position $\{\tau+h, x(\tau+h)\}$ is called the synthesis of the optimal controller.

It is obvious that the statement can be called dynamic.

From the given statement of the synthesis problem it follows that the processing of the controller begins from the known optimal program control for starting position $\{t=0, x(0)=x_0\}$. So using [19] we give the results necessary for the beginning of operation of the controller.

3.2.2. Optimal program.

Let $u(t)$, $t \geq 0$, be some control satisfying constraints (2.2) and bring the trajectory of the system (2.1) at the moment θ to zero: $x(\theta) = 0$ and also $u(\theta-h) = 0$.

Two cases are possible:

$$1)\ u(\theta-h) > 0$$

$$2)\ u(\theta-h) < 0.$$

Consider the first case for definiteness. We form the linear problem of optimal control

$$J(\theta) = u(\theta-h) \longrightarrow \min,$$

$$x(t+h) = A(h)x(t) + b(h)u(t),\ x(0)=x_0, \qquad (2.4)$$

$$x(\theta) = 0; \quad |u(t)| \le 1,\ t \in T_0(\theta-h)=\{0,h,\dots,\theta-2h\}.$$

Algorithms for solving the problems of type (2.4) are given in Chapter 1. We shall use these results.

If $u^0(t|t=0,x_0)$, $t\in T_0(\theta)$ is the optimal control, $T^0_{sup}(\theta)= \{\tau_1,\dots,\tau_n = \theta-h\} \in T_0(\theta)$ is an optimal support $(0 \le \tau_1<\dots< \tau_{n-1} < \tau_n = \theta-h)$ and $\min J(\theta) > 0$ then the totality $\{u^0(t|t=0,x_0),\ t \in T_0(\theta);\ T^0_{sup}(\theta);\ t^{*0}(x_0) = \theta\}$ is the solution to the time optimal control problem (2.1)-(2.3).

While solving problem (2.4) the value of $J(\theta)$ (the component of the control $u(\theta-h)$) is inspected. If this value becomes equal to zero at some iteration, then the process of the problem solution on $T_0(\theta)$ is stopped. The obtained information is used as the basis for the solution of the problem of type (2.4) where $T_0(\theta)$ is to substituted by $T_0(\theta-h)$.

Suppose that the solution

$$\{u^0(\cdot|t=0,x_0);\ T^0_{sup}(\theta);\ \theta\}$$

to problem (2.1)-(2.3) is obtained.

To the support $T^0_{sup}(\theta)=\{\tau_1,\dots,\tau_n = \theta-h\}$ correspond the matrices

$$P_{sup}=P_{sup}(\theta)=[A^{((\theta-\tau_i)/h)-1}b,\ i=\overline{1,n}],\ Q=Q(\theta)=P^{-1}_{sup}(\theta) \quad (2.5)$$

and the vector of potentials $\upsilon = q(\tau_n)$, where $q(\tau_n)$ is the row of the matrix Q corresponding to τ_n

The support $T^0_{sup}(\theta)$ is accompanied by the co-trajectory $\psi(t) = \psi(t|\theta)$, $t \in T_0(\theta)$, which is the solution to the conjugate system

$$\psi'(t-h)=\psi'(t)A(h), \psi(\theta-h)= -\upsilon.$$

It generates the co-control

$$\Delta(t)=\Delta(t|\theta), \quad t \in T_0(\theta): \quad \Delta(t)= -\psi'(t)b(h). \tag{2.6}$$

According to construction at the support moments the co-control (2.6) equals:

$$\Delta(t) = 0, \ t \in \{ \tau_1, \tau_2, \ldots, \tau_{n-1} \}, \ \Delta(\tau_n) = 1.$$

We shall find the optimal control at non-support moments using relations

$$u^0(t|t=0,x_0)\begin{cases} =u_*(t) \text{ when } \Delta(t)>0; \\ =u^*(t) \text{ when } \Delta(t)<0; \\ \in[u_*(t),u^*(t)] \text{ when } \Delta(t)=0, t\in T^0_N(\theta)=T_0(\theta)\backslash T_{sup}(\theta) \end{cases} \tag{2.7}$$

The totality $u^0_{sup}= (u^0(t),\ t \in T_{sup}(\theta))$ of values at support moments is calculated by

$$u^0_{sup} = u^0_{sup}(T^0_{sup}(\theta)) = - Q(A(h)^{\theta/h} x_0 - \sum_{t=t_*}^{t=\theta} A^{(\theta-t)/(h-1)} bu(t)$$

3.2.3. Optimal controller.

Let all the eigen-values of the matrix $A(h)$ be real. In this case at sufficiently small $h>0$ the sign of the co-control $\Delta(t|\theta)$, $t\in T_0(\theta)$, is changed not more than $n-1$ times, i.e. between support moments the co-control sign is constant.

Denote by

$$\{ u^0(t|\tau,x(\tau)),\ t\in T_\tau(\theta)=\{\tau,\ \tau+h, \ldots, \theta(\tau)-h\},\ T^\tau_{sup}(\theta(\tau));\theta(\tau)\}$$

a solution to problem (2.1)-(2.3) in the set $T_\tau(\theta)=\{\tau,\ \tau+h, \ldots, \theta(\tau)-h\}$. We suppose that the optimal control is unique for the optimal time $\theta(\tau)$ on $T_\tau(\theta)$. It can be constructed with the help of a formula of type (2.5)-(2.7). At support moments the control working off the perturbations $y(t,h)$ at $t<\tau$ equals

$$u^{o}_{sup} = u^{o}_{sup}(T^{o}_{sup}(\theta)|\tau,x(\tau)) = Q(\theta(\tau))g(\tau), \qquad (2.8)$$

$$g(\tau) = -A(h)^{\theta(\tau)/h}x_{o} - \sum_{t=0}^{\tau-h} A^{(\theta(\tau)-t)/h}y(t,h) -$$

$$- \sum_{t=0}^{\tau-h} A^{(\theta(\tau)-t)/(h-1)}bu(t|t,x(t)) -$$

$$- \sum_{t\in T^{\tau}_{N}(\theta(\tau))} A^{(\theta(\tau)-t)/(h-1)}bu(t|\tau,x(\tau)) ,$$

$$T^{\tau}_{N}(\theta) = T_{\tau}(\theta)\backslash T^{\tau}_{sup}(\theta).$$

We construct the sets

$$T^{\tau}_{N+}(\theta) = \{ t \in T^{\tau}_{N}(\theta):\Delta(t|\theta)>0, \ t-h \in T^{\tau}_{sup}(\tau)\},$$

$$T^{\tau}_{N-}(\theta) = \{t \in T^{\tau}_{N}(\theta):\Delta(t|\theta)<0, \ t-h \in T^{\tau}_{sup}(\theta)\}$$

and assume that the array

$$V_{\tau}(t|\theta) = A^{(\theta(\tau)-t)/(h-1)}b, \ t \in T^{\tau}_{N+}(\theta) \cup T^{\tau}_{N-}(\theta).$$

is known.

We describe now the optimal controller algorithm. In this context we shall consider that the system state does not leave the domain of controllability while the controller acts. The general case is investigated using the same scheme but it relies upon additional calculations.

We call the set

$$C^{k}(\tau) = \{ u^{(k)}(t), \ t \in T_{\tau+h}(\theta^{k}); \ \theta^{k}(\tau) ;$$

$$T^{(k)}_{sup} = \{ \tau_1, \ldots, \tau_n \};\ T^{(k)}_{N+};\ \ T^{(k)}_{N-};\ y^k;$$

$$V^k(t), t \in T^{(k)}_{N+} \cup T^{(k)}_{N-};\ \Delta^k(t), \Delta^k(t-2h),\ t \in T^{(k)}_{N+} \cup T^{(k)}_{N-}; Q^k \}$$

the state of the algorithm on the k-th iteration at τ. As the initial state $C^0(\tau)$ we choose the set with the following components:

$$u^{(0)}(t) = u^0(t|\tau, x(\tau)) \ ,\ t \in T_{\tau+h}(\theta(\tau));\ \theta^0(\tau) = \theta(\tau);$$

$$T^{(0)}_{sup} = T^{\tau}_{sup};\ T^{(0)}_{N+} = T^{\tau}_{N+};\ T^{(0)}_{N-} = T^{\tau}_{N-};$$

$$y^0 = \gamma(\tau, h);\ V^0(t) = V_\tau(t|\theta),\ t \in T^{\tau}_{N+} \cup T^{\tau}_{N-};$$

$$\Delta^0(t) = \Delta(t|\theta),\ \Delta^0(t-2h) = \Delta(t-2h|\theta), t \in T^{\tau}_{N+} \cup T^{\tau}_{N-};$$

$$Q^0 = Q(\theta(\tau)).$$

The algorithm iteration $C^k(\tau) \longrightarrow C^{k+1}(\tau)(C^k(\tau) \longrightarrow C^0(\tau+h))$ consists of the following steps.

Step 1. Compare τ with τ_1. If $\tau < \tau_1$ then pass to *Step 2*, for $\tau = \tau_1$ we pass to *Step 9*.

Step 2. Calculate the vector

$$\Delta u^k(T^{(k)}_{sup}) = (\Delta u^k(t), t \in T^{(k)}_{sup}) = -\ Q^k A^{(\theta^k - \tau)/h} y^k;$$

$$\Delta u^k(T^{(k)}_N) = 0, T^{(k)}_N = T_{\tau+h}(\theta^k) \setminus T^{(k)}_{sup}.$$

Step 3. Calculate the number μ^k and find the moment $\tau_s \in T^{(k)}_{sup}$:

$$\mu^k = \mu(\tau_s) = \min\ \mu(t), t \in T^{(k)}_{sup};$$

$$\mu(t)=\begin{cases}\dfrac{-1-u^{(k)}(t)}{\Delta u^{k}(t)}, & \text{when } \Delta u^{k}(t)<0;\\[2mm] \dfrac{1-u^{(k)}(t)}{\Delta u^{k}(t)}, & \text{when } \Delta u^{k}(t)>0;\\[2mm] \infty, & \text{in other cases}, t\in T^{(k)}_{sup}\setminus(\theta^{k}-h),\end{cases}$$

$$\mu(\theta^{k}-h)=\begin{cases}\dfrac{-u^{(k)}(\theta^{k}-h)}{\Delta u^{k}(t)}, & \text{when } u^{(k)}(\theta^{k}-h)\Delta u^{k}(\theta^{k}-h)<0;\\[2mm] \dfrac{1-|u^{(k)}(\theta^{k}-h)|}{|\Delta u^{k}(t)|}, & \text{when } u^{(k)}(\theta^{k}-h)\Delta u^{k}(\theta^{k}-h)>0;\\[2mm] \infty, & \text{when } \Delta u^{k}(\theta^{k}-h)=0.\end{cases}$$

When $\mu^{k}=1$ we pass to the next *Step*. In the case of $\mu^{k}<1$, $s=n$, we turn to *Step 5*. Let $\mu^{k}<1$, $s=n$, $\Delta u^{k}(\tau_{n})<0$. Then we pass to *Step 7*. In the case of $\mu^{k}<1$, $s=n$, $\Delta u^{k}(\tau_{n})>0$ we pass over *Step 8*.

Step 4. Pass to *Step 1* with the initial state $C^{0}(\tau+h)$ for $\tau+h$ with the following components:

$$u^{(0)}(t)=u^{(k)}(t)+\Delta u^{k}(t),\ t\in T_{\tau+2h}(\theta^{0}(\tau+h));$$

$$\theta^{0}(\tau+h)=\theta^{k}(\tau);\ T^{(0)}_{sup}=T^{k}_{sup};\ T^{(0)}_{N+}=T^{(k)}_{N+};\ T^{(0)}_{N-}=T^{(k)}_{N-};$$

$$y^{0}=y(\tau+h,h);\ \ V^{0}(t)=V^{k}(t),\ t\in T^{(0)}_{N+}\cup T^{(0)}_{N-};$$

$$\Delta^{0}(t)=\Delta^{k}(t),\ \Delta^{0}(t-2h)=\Delta^{k}(t-2h),\ t\in T^{(0)}_{N+}\cup T^{(0)}_{N-};$$

$$Q^{0}=Q^{k}.$$

Step 5. Calculate

$$\delta^{k}(t)=\rho q^{k}(\tau_{s})V^{k}(t),$$

$$\delta^{k}(t-2h)=\rho q^{k}(\tau_{s})A^{2}V^{k}(t),\ t\in T^{(k)}_{N+}\cup T^{(k)}_{N-},$$

$\rho=\operatorname{sign}\Delta u(\tau_{s})$ if $u^{K}(\tau_{n})>0$, $\rho=-\operatorname{sign}\Delta u(\tau_{s})$ if $u^{K}(\tau_{n})<0$

Find the numbers σ^{k}, t_{q}:

$$\sigma^k = \sigma(t_q) = min\ \{\sigma(t),\sigma(t-2h),\ t\in T^{(k)}_{N+}\cup T^{(k)}_{N-}\};$$

$$\sigma(t)=-\frac{\Delta^k(t)}{\delta^k(t)},\ \text{when}\ t\in T^{(k)}_{N+},\delta^k(t)<0\ \text{or}\ t\in T^{(k)}_{N-},\delta^k(t)>0;$$

$$\sigma(t)=-\frac{\Delta^k(t-2h)}{\delta^k(t-2h)},\ \text{when}\ t\in T^{(k)}_{N+},\ \delta^k(t-2h)<0\ \text{or}$$

$t\in T^{(k)}_{N-},\delta^k(t-2h)>0$, $\sigma(t)=\infty$, $\sigma(t-2h)=\infty$ in other cases.

For $\sigma^k=\infty$ pass to *Step* 8. Let $\sigma^k<\infty$. Then pass to the next *Step*.

Step 6. Let $T^{(k+1)}_{sup} = (T^{(k)}_{sup}\backslash\tau_s)\cup t_q$,

$$Q^{k+1}(\tau_i,j)=Q^k(\tau_i,j)-Q^k(\tau_s,j)r^k(\tau_i)/r^k(\tau_s),\ i=s;$$

$$Q^{k+1}(\tau_s,j)=Q^k(\tau_s,j)/r^k(\tau_s),\ i=\overline{1,m},\ j=\overline{1,m} \tag{2.9}$$

where $r^k=(r^k(\tau_i),i=\overline{1,m})=Q^kV^k(t_q)$, when $t_q\in T^{(k)}_{N+}\cup T^{(k)}_{N-}$,

$$r^k=Q^kA^2V^k(t_q),\ \text{when}\ t_q\in T^{(k)}_{N+}\cup T^{(k)}_{N-};$$

$$u^{(k+1)}(t)=u^{(k)}(t)+\mu^k\Delta u^k(t),\ t\in T_{\tau+h}\ (\theta^k);$$

$$\theta^{k+1}(\tau)=\theta^k(\tau);\ y^{k+1}=(1-\mu^k)y^k;$$

$$\Delta^{k+1}(t)=\Delta^k(t)+\sigma^k\delta^k(t),\ \Delta^{k+1}(t-2h) = \Delta^k(t-2h)+\sigma^k\delta^k(t-h),$$

$$t\in T^{(k)}_{N+}\cup T^{(k)}_{N-},\ \Delta^{k+1}(t_q+h) = q^{k+1}(\tau_n)A^{-1}V^k(t_q),$$

$$\Delta^{k+1}(t_q-h) = q^{k+1}(\tau_n)AV^k(t_q),\ t_q\in T^{(k)}_{N+}\cup T^{(k)}_{N-};$$

$$\Delta^{k+1}(t_q+h) = q^{k+1}(\tau_n)AV^k(t_q),$$

$$\Delta^{k+1}(t_q-h) = q^{k+1}(\tau_n)A^{-1}V^k(t_q), \; t_q \in T^{(k)}_{N+} \cup T^{(k)}_{N-}.$$

Exclude the moment τ_s+h from $T^{(k)}_{N+} \cup T^{(k)}_{N-}$ and add accordingly the moment t_q+h in $T^{(k+1)}_{N+}$, when $\Delta^{k+1}(t_q+h) > 0$ or in $T^{(k+1)}_{N-}$ when $\Delta^{k+1}(t_q+h) < 0$:

$$T^{(k+1)}_{N+} \cup T^{(k+1)}_{N-} = (T^{(k)}_{N+} \cup T^{(k)}_{N-}) \setminus (\tau_s+h) \cup (t_q+h);$$

$$V^{k+1}(t) = V^k(t), \; t \in (T^{(k+1)}_{N+} \cup T^{(k+1)}_{N-}) \setminus (t_q+h),$$

$$V^{k+1}(t_q+h) = A^{-1}V^k(t_q).$$

Pass to *Step* 2.

Step 7. If $\theta^k-\tau > n$, let $\theta^{k+1}=\theta^k-h$, $Q^{k+1}= \tilde{Q}^kA$, $\tilde{Q}^k$ calculate using the reccurent formulas (2.9) ($\tau_s=\tau_n$, $t_q = \tau_n - k_0h$, $k_0 = \min k$, $\tau_n - kh \in T^{(k)}_N$). Calculate

$$u^{(k+1)}(t)=u^{(k)}(t)+\mu^k\Delta u^k(t), t\in T_{\tau+h}(\theta^{k+1}),$$

$$V^{k+1} = A^{-1}V^k(t), \; t \in T^{(k)}_{N+} \cup T^{(k)}_{N-}, \; y^{k+1}=(1-\mu^k)y^k,$$

and form the sets

$$T^{(k+1)}_{N+} \cup T^{(k+1)}_{N-} = (T^{(k)}_{N+} \cup T^{(k)}) \setminus t_q \;, \text{ if } t_q \in T^{(k)}_{N+} \cup T^{(k)}_{N-},$$

$$T^{(k+1)}_{N+} \cup T^{(k+1)}_{N-} = (T^{(k)}_{N+} \cup T^{(k)}) \;, \text{ if } t_q \notin T^{(k)}_{N+} \cup T^{(k)}_{N-},$$

$$T^{(k+1)}_{sup} = (T^{(k)}_{sup} \setminus \tau_s) \cup t_q.$$

If $k_0=1$ then

$$\Delta^{k+1}(t) = \Delta^k(t)/\Delta^k(t_q), \; \Delta^{k+1}(t-2h) = \Delta^k(t-2h)/\Delta^k(t_q),$$

$$t\in T^{(k)}_{N+} \cup T^{(k)}_{N-}.$$

For $k_0 > 1$

$$\Delta^{k+1}(t) = q^{k+1}(\tau_{n-1})/ V^k(t) - (\Delta^k(t)A^2V^k(t_q))/\Delta^k(t_q),$$

$$\Delta^{k+1}(t-2h) = q^{k+1}(\tau_{n-1})/ V^k(t) - (\Delta^k(t-2h)A^2V^k(t_q))/\Delta^k(t_q),$$

$$t \in T^{(k)}_{N+} \cup T^{(k)}_{N-}.$$

Pass to *Step 2*.

For $\theta^k - \tau = n$ we find

$\tilde{\mu}(\tau_n) = 1/ |\Delta u(\tau_n)|$, $\tilde{\mu}^k = \tilde{\mu}(\tau_s) = \min \{ \mu(\tau_1), \ldots ,\mu(\tau_{n-1}), \tilde{\mu}(\tau_n)\}$. If $\tilde{\mu}^k > 1$, then pass to *Step 4*. If $\tilde{\mu}^k < 1$ then pass to *Step 8*.

Step 8. Let $\theta^{k+1} = \theta^k + h$, $\tilde{Q}^k = Q^k A^{-1}$; Q^{k+1} is obtained from $\tilde{Q}^k$ according to the reccurent formula (2.9) substituting moment τ_s for $t_q = \theta^k$ $(T^{(k+1)}_{sup} = (T^{(k)}_{sup} \backslash \tau_s) \cup t_q)$. Calculate the estimates

$$\Delta^{k+1}(t) = \Delta^k(t)/q^k(\tau_s)A^{-1}b,$$

$$\Delta^{k+1}(t-2h) = \Delta^k(t-2h)/q^k(\tau_s)A^{-1}b, \text{ if } s \neq n;$$

$$\Delta^{k+1}(t) = \delta^k(t)/\rho q^k(\tau_s)A^{-1}b,$$

$$\Delta^{k+1}(t-2h) = \delta^k(t-2h)/\rho q^k(\tau_s)A^{-1}b, \text{ if } s = n;$$

$$t \in T^{(k)}_{N+} \cup T^{(k)}_{N-},$$

and form the sets

$$T^{(k+1)}_{N+} \cup T^{(k+1)}_{N-} = T^{(k)}_{N+} \cup T^{(k)} \cup \tau_s,$$

$$(\Delta^{k+1}(t) = 1/q^k(\tau_s)A^{-1}b) \text{ ,if } \tau_s - h \in T^{(k+1)}_{sup};$$

$$T^{(k+1)}_{N+} \cup T^{(k+1)}_{N-} = T^{(k)}_{N+} \cup T^{(k)}_{N-}, \text{ if } \tau_s - h \notin T^{(k+1)}_{sup}.$$

Let

$$u^{(k+1)}(t)=u^{(k)}(t)+\mu^k \Delta u^k(t), t \in T_{\tau+h}(\theta^k),$$

$$V^{k+1} = AV^k(t), \ t \in T^{(k+1)}_{N+} \cup T^{(k+1)}_{N-}; y^{k+1}=(1-\mu^k)y^k,$$

$$u^{(k+1)}(\theta^k)=0.$$

Pass to *Step 5*.

Step 9. Let $s=1$, $\mu^0=0$, $\Delta u^0(\tau_1) = -\ q^0(\tau_1)A^{(\theta^0(\tau)-\tau)/h}\ y^0$, and pass to *Step 5*.

Example 2.1. Let us illustrate the results described above using the example of the discrete control system

$$t^* \longrightarrow max, \ x_1(t+h) = x_1(t) + hx_2(t), \ x_2(t+h) = x_2(t)+hu(t),$$

$$x_1(0) = 4, \ x_2(0)=-2, \ x(t^*) = 0, \ 0 \le u(t) \le 1, t = 0,1,\dots,h=1.$$

The optimal program control $u^0(t|\ t= 0,x_0)$ is represented in Fig. 2.1.

The optimal control $u^0(t|\ t= 1,x_1)$ produced by the controller for parrying the perturbation $y_1(0) = 0$, $y_2(0) = -1/2$, is given in Fig. 2.2. It was obtained from the optimal program control for 5 iterations.

Intermediate values of the optimal control corresponding to parrying parts of the perturbation take the form given in Figs. 2.3-2.6. It should be stressed that the treated perturbation increased the optimal time $(t^{*0}(0,x_0)= 3$, $t^{*0}(1,x_1) = 5)$.

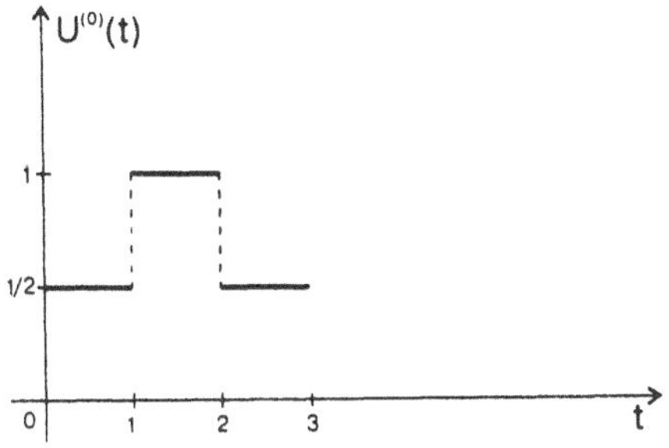

Fig.2.1

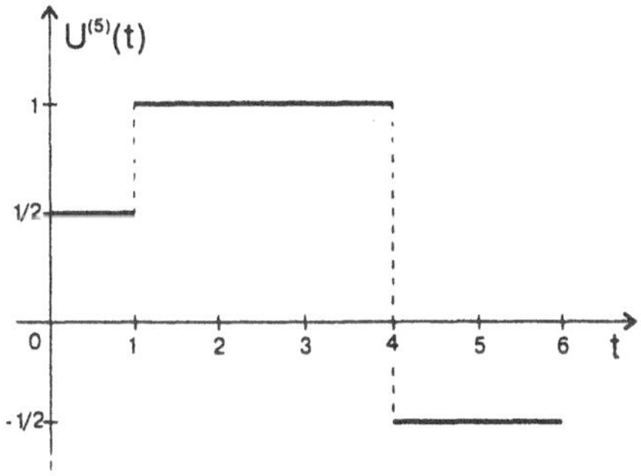

Fig.2.2

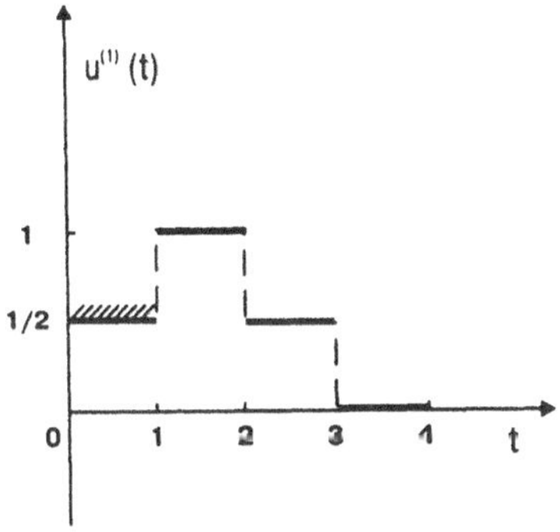

Fig.2.3

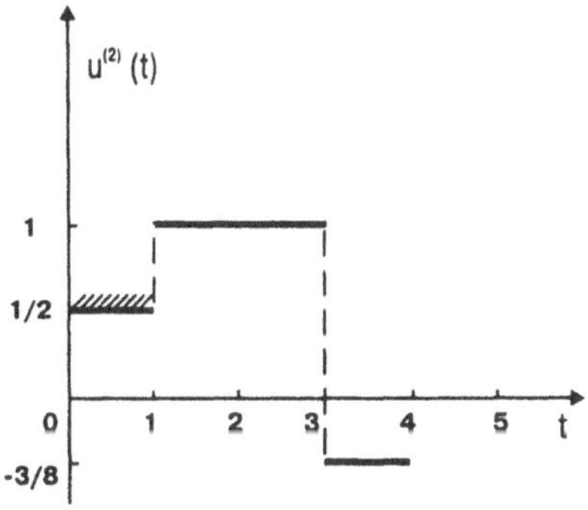

Fig.2.4

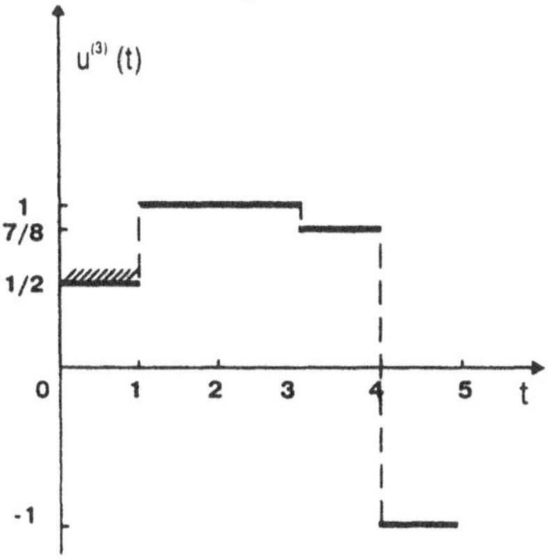

Fig.2.5

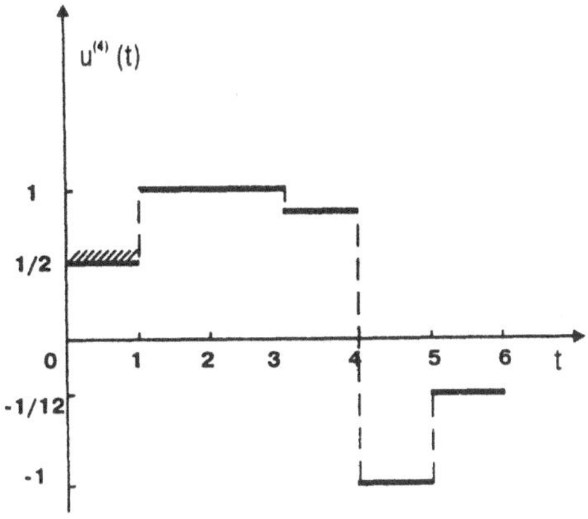

Fig.2. 6

3.3. OPTIMAL CONTROLLER FOR DISCRETE SYSTEMS WITH INDETERMINATE PERTURBATIONS.

Under optimal system synthesis the controller algorithm depends on the accessible information about perturbations acting on the system. In Section 3.1 we described the controller counteracting perturbations which are measured during the control process. Now we consider another formulation of the synthesis problem. We shall assume that under control design a perturbation is unknown and only the domain of its possible values is known. Since the quantization period is assumed to be small and is defined by the controller speed the variation of the system states over this period will not be large. In this context it is interesting to establish a relation between the results obtained in this section and Section 3.1.

3.3.1. Statement of the problem.

Consider the terminal problem of optimal control

$$J(u) = h_0'x(t^*) \longrightarrow max,$$

$$x(t+h) = A(h)x(t) + b(h)u(t), x(0)=x_0, \qquad (3.1)$$

$$h_i'x(t^*) \geq g_i, \; i=\overline{1,m};$$

$$u_* \leq u(t) \leq u^*(t),$$

$$t \in T(0)=\{0,h,\dots,t^*-h\}.$$

As usual we shall distinguish between two types of solution to problem (3.1): program and feedback.

The algorithm of control design of one type of feedback is suggested in Section 3.1. The aim of this section is to describe the algorithm of design of another type of optimal feedback. Its difference from the one considered earlier (see Section 3.1) is connected with a new kind of accessible information about perturbations.

As in Section 3.1 the controller begins to operate at the moment $t=0$ from the state $x(0)=x_0$ and its work is based on a special solution of problem (3.1). Let the controller operate during

$$t=0,h,\ldots,\tau-h,$$

and the system at the moment $t=\tau$ under the action of controls

$$u(0),\ldots,u(\tau-h),$$

and the perturbations

$$y(0,h),\ldots,y(\tau-h,h)$$

be in the state $x(\tau)$. Now unlike Section 3.1, at the moment τ it is not known what perturbation $y(\tau,h)$ will act on the state $x(\tau)$. It is only known that

$$y(\tau,h) \in Y = \{ z \in R^n: Gz = f,\ d_* \le z \le d^* \}. \tag{3.2}$$

We shall call controller synthesis (or, in other words by control design of feedback type) the calculation of the control

$$u^0(t|\tau,x(\tau)),\ t=\tau,\ \tau+h,\ldots,t^*-h,$$

by which all the trajectories of system (3.1) with initial conditions

$$x(\tau) + y(\tau,h),\ y(\tau,h) \in Y,$$

will satisfy the terminal restrictions

$$h_i'x(t^*) \ge g_i,\ i=1,m,$$

and the value of the quality criterion

$$J(u(\cdot)) = \min_{y(\cdot)\in Y} h_0'x(t^*)$$

will be a maximum ($J(u^o(\cdot)) = \max J(u(\cdot))$).

3.3.2. Program control under uncertainty conditions.

According to Section 3.3.1 the result

$$u^o(t|\tau,x(\tau)),\ t=\tau,\tau+h,\dots,t^*-h,$$

of the synthesis at the moment τ is a solution of the following extremal problem :

$$J(u) = h_o'x(t^*) \longrightarrow \max,$$

$$x(t+h) = A(h)x(t) + b(h)u(t), x(\tau)=x_\tau,$$

$$h_i'x(t^*) \geq g_i^\tau,\ i=\overline{1,m}; \qquad (3.3)$$

$$u_* \leq u(t) \leq u^*(t),\ t \in T(\tau),$$

where x_τ is the measured state $x(\tau)$ at τ;
$g_i^\tau = g_i - \gamma_i^\tau$, γ_i^τ is an estimate of the following linear programming problem:

$$\gamma_i^\tau = \min h_i'A^{(t^*-\tau)/h}z, \qquad (3.4)$$

$$Gz = f,\ d_* \leq z \leq d^*.$$

The solution to the terminal control problem (3.3) is the set

$$\{u^o(\cdot|\tau,x(\tau)),\ S_{sup}(\tau)\},$$

where

$$S_{sup}(\tau)= \{ I_{sup}(\tau),\ T_{sup}(\tau)\},\ I_{sup}(\tau) \subset I = \{ 1,2,\dots,m\},$$

$$T_{sup}(\tau) = \{\tau_1,\dots,\tau_l\},\ \tau \leq \tau_1(\tau) < \tau_2(\tau) < \dots<\tau_l(\tau)\leq t^*-h.$$

The relations

$$|I_{sup}(\tau)|=|T_{sup}(\tau)|=l,\ 0 \le l \le m,\ \det P(\tau) \ne 0,$$

$$P(\tau)=\begin{bmatrix} h_i'A^{(t^*-t)/h-1}b,\ t\in T_{sup}(\tau) \\ i \in I_{sup}(\tau) \end{bmatrix}$$

are fulfilled.

The vector of potentials corresponds to the support

$$\nu' = \nu'(\tau) = c_{sup}'Q(\tau),$$

$$c_{sup}=(c(t),t\in T_{sup}(\tau)),\quad c(t) = h_0'A^{(t^*-t)/h-1}b,$$

$$\tau \le t \le t^*-h,\ Q(\tau) = P^{-1}(\tau),$$

and with the help of which the accompanying co-trajectory

$$\psi(t) = \psi(t|\tau),\ \tau \le t \le t^*-h$$

is constructed as a solution to the conjugated system

$$\psi'(t-h) = \psi'(t)A(h),\quad \psi'(t^*-h) = h_0'-\upsilon'(I_{sup})H(I_{sup},J),$$

$$H(I,J) = \begin{bmatrix} h_i'(J) \\ i\in I \end{bmatrix}.$$

The co-trajectory generates the co-control

$$\Delta(t)=\Delta(t|\tau),\ \tau \le t \le t^*-h:\ \Delta(t)= -\psi'(t)b(h). \tag{3.5}$$

The optimal control $u^0(t)$, $t \in T_N(\tau) = T(\tau)\backslash T_{sup}(\tau)$, at the non-support moments of time is calculated by (1.9).

Without loss of generality we may consider that

$$h_i'x(t^*) = g_i^\tau,\ i \in I_{sup}(\tau)\ (\ \upsilon(i) \le 0, i \in I_{sup}(\tau)\ ).$$

The set of values $u^{o}_{sup} = (u^{o}(t),\ t \in T_{sup}(\tau)\ )$ of optimal control at the support moments is found by

$$u^{o}_{sup} = Q(\tau)g(\tau),$$

where

$$g(\tau)=\begin{bmatrix} g_i(\tau) \\ i \in I_{sup}(\tau) \end{bmatrix},\ g_i(\tau) = g_i^{\tau} -$$

$$-\sum_{t\in T_N(\tau)} (h_i{}'A^{(t^{*}-t)/h-1}bu(t)- h_i'A^{(t^{*}-\tau)/h}x_{\tau}.$$

We construct the sets $T_{N+}(\tau)$, $T_{N-}(\tau)$ according to (1.11) and suggest

1) $|T_{N+}(\tau)| + |T_{N-}(\tau)| = m + 1$, $(\ \tau_N \in T_{N+}(\tau) \cup T_{N-}(\tau),\ \tau_N - h \in$ $\in T_{sup}(\tau)\ )$.

2) the information array is known

$$V_{\tau}(t) = A^{(t^{*}-t)/h-1}b,\ t \in T_{sup}(\tau) \cup \tau_N \cup (t^{*}-h).$$

3.3.3. Optimal controller synthesis.

As the state of the algorithm

$$C^{k}(\tau) = \{u^{(k)}(t), t\in T(\tau);\ W^{k};\ S^{k}_{sup};T^{k}_{N+};T^{k}_{N-};$$

$$\Delta g^{k};\ V^{k}(t), t \in T^{k}_{sup} \cup \tau_N \cup (t^{*}-h);\ \Delta^{k}(t-h), \Delta^{k}(t+h),\ t \in T^{k}_{sup},$$

$$\Delta(t),\ t=\tau_N, \tau_N-h, t^{*}-h;\ Q^{k};\ \upsilon^{k}\}$$

at the zeroth-iteration at the moment τ let us choose the set with the following components:

$$u^{(0)}(t)=u^0(t|\tau-h),\ t\in T(\tau);\ W^0=Hx^0(t^*)-g^\tau;$$

$$S^0_{sup}=S_{sup}(\tau);\ T^0_{N+}=T_{N+}(\tau-h);\ T^0_{N-}=T_{N-}(\tau-h);$$

$$\Delta g^0=\gamma^{\tau-h}-\gamma^\tau+H(Ax_{\tau-h}+bu^0(\tau-h|\tau-h,x(\tau-h))-x_\tau);$$

$$V^0(t)=V_\tau(t-h),\ t\in T^0_{sup}\cup\tau_N\cup(t^*-h);$$

$$\Delta^0(t-h)=\Delta(t-h|\tau-h),\ \Delta^0(t+h)=\Delta(t+h|\tau-h),t\in T^{\ 0}_{sup};$$

$$\Delta^0(\tau_N)=\Delta(\tau_N|\tau-h),\ \Delta^0(\tau_N-h)=\Delta(\tau_N-h|\tau-h),$$

$$\Delta^0(t^*-h)=\Delta(t^*-h|\tau-h)\ ;\ Q^0=Q(\tau-h);\ \upsilon^0=\upsilon(\tau-h).$$

The algorithm iteration $C^k(\tau)\longrightarrow C^{k+1}(\tau)(C^k(\tau)\longrightarrow C^0(\tau+h))$ consists of the following steps.

Step 1. If $l=0$, then pass to *Step 2*. Let $l\geq 1$. Compare $\tau-h$ with τ_1. If $\tau-h<\tau_1$ then pass to the following *Step*. At $\tau-h=\tau_1$ pass to *Step 8*.

Step 2. Calculate the vectors

$$\Delta u^k(T^{\ k}_{sup})=(\Delta u^k(t),t\in T^{\ k}_{sup})=Q^k\Delta g^k(I^{\ k}_{sup});$$

$$\Delta u^k(T^k_N)=0,T^k_N=T(\tau)\setminus T^{\ k}_{sup};$$

$$\Delta W^k(I^k_N)=\sum_{t\in T^k_{sup}}H(I^k_N,J)V^k(t)\Delta u^k(t),$$

$$\Delta W^k(I^{\ k}_{sup})=0,\ I^k_N=I\setminus I^{\ k}_{sup}.$$

Step 3. Calculate the numbers $\alpha^k, \beta^k, \theta^k$:

$$\alpha^k=\alpha(\tau_s)=\min \alpha(t),\ t\in T^k_{sup};$$

$$\alpha(t)=\begin{cases} -\dfrac{u_*(t)-u^{(k)}(t)}{\Delta u^k(t)}, & \text{when} \quad \Delta u^k(t)<0; \\ \dfrac{u^*(t)-u^{(k)}(t)}{\Delta u^k(t)}, & \text{when} \quad \Delta u^k(t)>0; \\ \infty, & \text{when} \quad \Delta u^k(t)=0,\ t\in T^k_{sup}. \end{cases}$$

$$\beta^k=\beta(i_0)= \min \beta(i),\ i\in I^k_N;$$

$$\beta(i)=\begin{cases} -\dfrac{W^k_i}{\Delta W^k_i - \Delta g^k_i}, & \text{when } \Delta W^k_i - \Delta g^k_i < 0; \\ \infty, \text{ when} & \Delta W^k_i - \Delta g^k_i \ge 0. \end{cases}$$

Let $\theta^k= \min \{1, \alpha^k, \beta^k\}$. If $\theta^k = 1$, then pass to *Step* 4. At $\theta^k < 1$ we pass to *Step* 5.

Step 4. Assume $u^0(\tau|\tau,x(\tau)) = u^{(k)}(\tau) + \Delta u^k(\tau)$. If $\tau = t^*-lh$ then the algorithm completes the work:

$$u^0(\tau+ih|\tau+ih,x(\tau+ih)) = u^{(k)}(\tau+ih)+\Delta u^k(\tau+ih),\ i=\overline{1,l-1}.$$

If $\tau < t^*-lh$ we construct the initial state $C^0(\tau+h)$ for the moment $\tau+h$ with the following components:

$$u^{(0)}(t)=u^{(k)}(t)+\Delta u^k(t),\ t\in T(\tau+h);\ W^0= W^k+\Delta W^k;$$

$$S^0 =S^k_{sup};\ T^0_{N+}=T^k_{N+};\ T^0_{N-}=T^k_{N-};$$

$$\Delta g^0=\gamma^\tau-\gamma^{\tau+h}+H(Ax_\tau+bu^0(\tau|\tau,x(\tau))-x_{\tau+h});$$

$$\Delta^0(t-h)=\Delta^k(t-h),\ \Delta^0(t+h)=\Delta^k(t+h),\ t\in T^0_{sup};$$

$$\Delta^0(\tau_N) = \Delta^k(\tau_N),\ \Delta^0(\tau_N-h) = \Delta^k(\tau_N-h),\Delta^0(t^*-h) = \Delta^k(t^*-h);$$

$$V^o(t)=V^k(t), t\in T^o_{sup} \cup \tau_N \cup (t^*-h); \; Q^o=Q^k \; ; \; \upsilon^o=\upsilon^k.$$

Pass to *Step 1*.

Step 5. If $\theta^k=\beta^k=\beta(i_o)$ calculate

$$\mu^k(I^k_{sup}) = [h'_{i_o} V^k(t), \; t\in T^k_{sup}]Q^k,$$

$$\delta^k(t+h) = [\; h'_{i_o} - \mu^{k'}(I^k_{sup})H(I^k_{sup},J)]A^{-1}V^k(t), \; t\in T^k_{sup};$$

$$\delta^k(t-h) = [\; h'_{i_o} - \mu^{k'}(I^k_{sup})H(I^k_{sup},J)]AV^k(t), \; t\in T^k_{sup}\cup\tau_N;$$

$$\delta^k(t) = [\; h'_{i_o} - \mu^{k'}(I^k_{sup})H(I^k_{sup},J)]V^k(t), \; t\in\{\tau_N, t^*-h\}.$$

Provided $\theta^k=\alpha^k=\alpha(\tau_s)$ calculate

$$\mu^k(I^k_{sup}) = \rho q^k(\tau_s),$$

$$\delta^k(t+h) = \rho q^k(\tau_s)H(I^k_{sup},J)]A^{-1}V^k(t), \; t\in T^k_{sup};$$

$$\delta^k(t-h) = \rho q^k(\tau_s)H(I^k_{sup},J)]AV^k(t), \; t\in T^k_{sup}\cup\tau_N;$$

$$\delta^k(t) = \rho q^k(\tau_s)H(I^k_{sup},J)]V^k(t), \; t\in\{\tau_N, t^*-h\}.$$

Pass to *Step 6*.

Step 6. Calculate the numbers

$$\sigma^k= \; min \; \{ \; \sigma(t-h), \; \sigma(t+h), \; t \in T^k_{sup}; \sigma(t), \; t=\tau_N, \; \tau_N-h, \; t^*-h;$$

$$\omega(i), \; i\in I^k_{sup}\}; \; s(t), t\in T^k_{N+}\cup T^k_{N-}:$$

$$\sigma(t-h)=-\frac{\Delta^k(t-h)}{\delta^k(t-h)}, s(t)=h, \text{ when } t-h\in T^k_{N+}, \delta^k(t-h)<0 \text{ or } t-h\in T^k_{N-},$$

$\delta^k(t)>0$; $\sigma(t+h)=-\dfrac{\Delta^k(t+h)}{\delta^k(t+h)}$, $s(t)=-h$, when $t+h\in T^k_{N+}$, $\delta^k(t-h)<0$ or

$t+h\in T^k_{N-}$, $\delta^k(t-h)>0$, $t\in T^k_{sup}$; $\sigma(t)=-\dfrac{\Delta^k(t)}{\delta^k(t)}$, when $\Delta^k(t)\delta^k(t)<0$,

$t=\tau_N$, τ_N-h, t^*-h, $s(t)=0$; $\sigma(t-h)=\sigma(t+h)=\sigma(t)=\infty$ in other cases.

$\omega(i)= -\dfrac{\upsilon^k(i)}{\mu^k(i)}$, when $\upsilon^k(i)\mu^k(i) < 0$ or $\upsilon^k(i)=0$, $\mu^k(i) > 0$,

$\omega(i)=\infty$ in other cases, $i \in I^k_{sup}$. Pass to *Step 7*.

Step 7. Transform the set $S^k_{sup} = \{I^k_{sup}, T^k_{sup}\}$ and the matrix Q^k.

1) Let $\theta^k = \beta(i_0) < 1$, $\sigma^k = \omega(i_*)$. Then

$$I^{k+1}_{sup} = (I^k_{sup}\setminus i_*) \cup i_0,\ T^{k+1}_{sup} = T^k_{sup},$$

$$Q^{k+1}(\tau_j,i) = Q^k(\tau_j,i) + Q^k(\tau_j,i_*)r^k(i)/r^k(i_*), i=i_*;$$

$$Q^{k+1}(\tau_j,i_*)=Q^k(\tau_j,i_*)/r^k(i_*), j=\overline{1,l}, i=\overline{1,l},$$

where $r^k=(r^k(i), i\in I^k_{sup}) = [h'_{i_0} V^k(t),\ t\in T^k_{sup}]Q^k(T^k_{sup}, I^k_{sup})$.

2) Let $\theta^k = \beta(i_0) < 1$, $\sigma^k = \sigma(t_q-s(t_q))$. Then

$$I^{k+1}_{sup} = I^k_{sup} \cup i_0,\ T^{k+1}_{sup} = T^k_{sup} \cup (t_q-s(t_q))$$

$$Q^{k+1}(\tau_j,i) = Q^k(\tau_j,i) + r^k_1(\tau_j)r^k_2(i)/\rho\delta(t_q-s(t_q)),$$

$$Q^{k+1}(t_q-s(t_q),i) = -r^k_2(i)/\rho\delta(t_q-s(t_q)),$$

$$Q^{k+1}(\tau_j,i_0) = -\ r^k_1(\tau_j)/\rho\delta(t_q-s(t_q)),$$

$$Q^{k+1}(t_q-s(t_q),i_0) = -1/\rho\delta(t_q-s(t_q)), j = \overline{1,l}, i = \overline{1,l},$$

$$r_1^k=(r_1^k(\tau_j),j=\overline{1,l}) = Q^k H(I_{sup}^k,J)V^k(t_q-s(t_q))),$$

$$r_2^k=(r_2^k(i),i=\overline{1,l}) = [h_{i_0}' V^k(t),\ t\in T_{sup}^k]Q^k.$$

3) Let $\theta^k = \alpha^k = \alpha(\tau_s) < 1$, $\sigma^k = \omega(i_*)$. Then

$$I_{sup}^{k+1} = I_{sup}^k \setminus i_*,\quad T_{sup}^{k+1} = T_{sup}^k \setminus \tau_s,$$

$$Q^{k+1}(T_{sup}^{k+1},I_{sup}^{k+1}) = Q^k(T_{sup}^k\setminus\tau_s,I_{sup}^k\setminus i_*)-$$

$$-Q^k(T_{sup}^k\setminus\tau_s,i_*)Q^k(\tau_s,I_{sup}^k\setminus i_*)/\rho\mu(i_*).$$

4) Let $\theta^k = \alpha^k = \alpha(\tau_s) < 1$, $\sigma^k = \sigma(t_q-s(t_q))$. Then

$$I_{sup}^{k+1} = I_{sup}^k,\quad T_{sup}^{k+1} = (T_{sup}^k\setminus\tau_s) \cup (t_q-s(t_q)),$$

$$Q^{k+1}(\tau_j,i) = Q^k(\tau_j,i) - Q^k(\tau_s,i)r^k(\tau_j)/r^k(\tau_s),i=i_*;$$

$$Q^{k+1}(\tau_s,i)=Q^k(\tau_s,i)/r^k(\tau_s),\ j=\overline{1,l},i=\overline{1,l},$$

$$r^k=(r^k(\tau_j),j=\overline{1,l}) = Q^k H(I_{sup}^k,J)V^k(t_q-s(t_q)).$$

The vector $V^k(t_q-s(t_q))$ is calculated in the standard way with respect to the vector $V^k(t_q)$.

Let

$$u^{(k+1)}(t)=u^{(k)}(t)+\theta^k\Delta u^k(t),t\in T(\tau);$$

$$\Delta g^{k+1}=(1-\theta^k)\Delta g^k;\ \Delta^{k+1}(t)=\Delta^k(t)+\sigma^k\delta^k(t),\ t\in T_{N+}^k\cup T_{N-}^k\cup(t^*-h);$$

$$\upsilon^{k+1} = \upsilon^k + \sigma^k\mu^k;\ W^{k+1} = W^k + \theta^k\Delta W^k.$$

Let in Cases 1),3) $T^{k+1}_{N+} \cup T^{k+1}_{N-} = (T^{k}_{N+} \cup T^{k}_{N-})$, in Case 2) $(T^{k+1}_{N+} \cup T^{k+1}_{N-}) = (T^{k}_{N+} \cup T^{k}_{N-}) \cup (t_q - s(t_q)+h)$ and in Case 4) $(T^{k+1}_{N+} \cup \cup T^{k+1}_{N-}) = (T^{k}_{N+} \cup T^{k}_{N-}) \setminus (\tau_s+h) \cup (t_q-s(t_q)+h)$.

Pass to *Step 2*.

Step 8. Let $s=1$, $\theta^0=\alpha^0=\alpha(\tau_1)=0$, $\Delta u^0(\tau_1)=q(\tau_1)\Delta g(I^{k}_{sup})$, and pass to *Step 5*.

Remarks:

1) It is not difficult to test that the synthesis algorithm is extended to include the case of nonstationary domain of perturbations when the set $Y = Y(x_\tau,\tau)$ depends both on time and values of the current state.

2) The algorithm is obviously generalized for the case when it is known that the perturbation will act not only at the moment τ but at the moment $\tau + h,\dots,\tau + rh \le t^* - h$.

Example 3.1. Let us illustrate the obtained results given above by the simple example of the mechanical motion optimization.

The material point, starting the motion in the rectilinear way from some neighbourhood of the given point and being acted upon by indeterminate perturbations, is required to be moved into the given domain at the given moment and to acquirethe velocity, the guaranteed value of which is a maximum.

The mathematical model of the problem is

$$x_2(3) \longrightarrow max, \; x_1(t+h)=x_1(t)+hx_2(t),$$

$$x_2(t+h)=x_2(t)+hu(t), x_1(3) \le 2, \; x_1(0) = x_2(0) = 0,$$

$$0 \le u(t) \le 1, \; -1 \le y_1(t) \le 1,$$

$$y_2(t) = 0, \; t = 0,h,\dots,3, \; h = 0.5.$$

Let us present the results of the optimal controller operation for the case when the perturbations

$$y_1(0) = 1/2,\ y_1(0.5) = 1/4,\ y_1(1) = -1/2,$$

$$y_1(1.5) = -1/4,\ y_1(2) = 1/4,\ y_1(2.5) = 0,$$

were actually realized but the values of this noise are unknown for the controller at the corresponding moments

During the operation described above the controller has designed the control presented in Fig. 3.1.

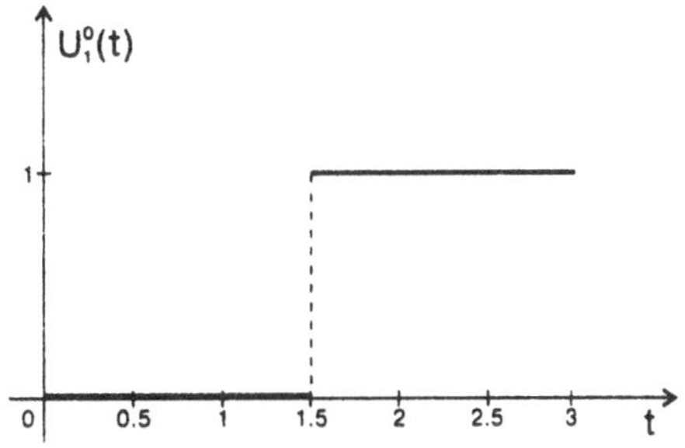

Fig.3.1

The efficiency of this control is equal to $J(u_1^o(\cdot)) = 3/2$.

If the information about the perturbations entered during the process the controller would design the control presented in Fig. 3.2.

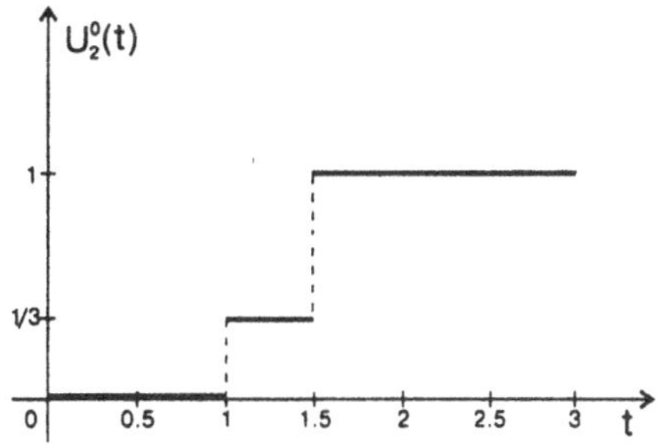

Fig.3.2

The value of the criterion for the designed control is equal to $J(u_2^o(\cdot)) = 5/3$.

Let the perturbation be known before the beginning of process. Then the optimal control has the configuration presented in Fig. 3.3. The effectiveness of the control is equal to $J(u_3^o(\cdot)) = 9/4$.

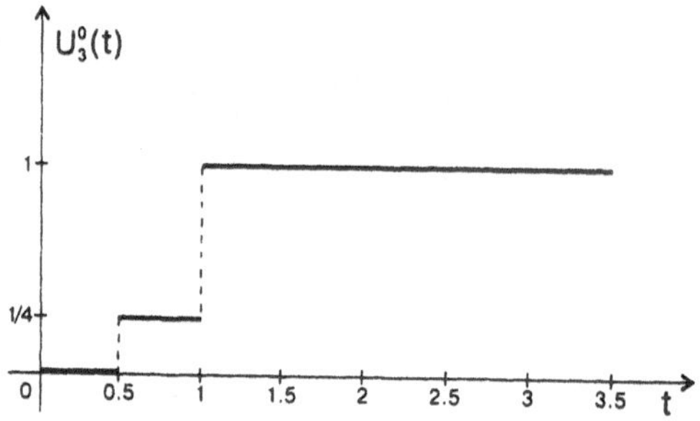

Fig.3.3

The controller from Section 3.1 produces the control for the nearest period after the perturbation measurement which is made in the present period, already after the adopted control. Assume that the perturbation measurement in each period is made before the control selection. In this case the controller produces the control presented in Fig.3.4. The algorithm for its operation is obtained according to a scheme close to the one given in Section 3.1 ($J(u_4^o(\cdot)) = 2$).

From the above example the dependence between the control efficiency and the prevailing information conditions is seen. In relation to full information the loss of efficiency is equal to $J(u_3^o(\cdot)) - J(u_1^o(\cdot)) = 3/4$, in relation to partial information it is equal to $J(u_2^o(\cdot)) - J(u_1^o(\cdot)) = 1/6$.

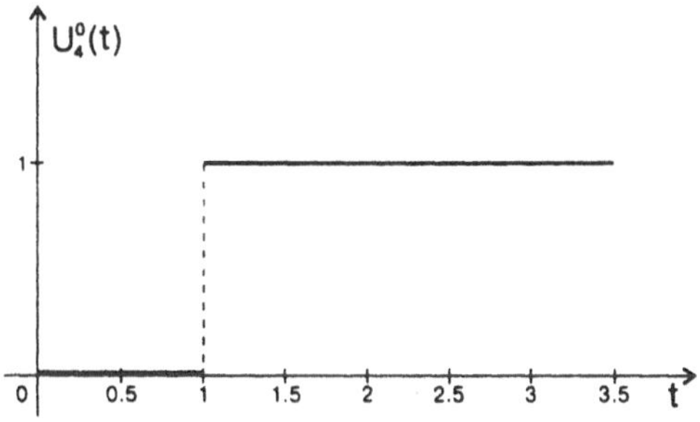

Fig.3.4

3.4. OPTIMAL CONTROLLER WITH PERTURBATION PREDICTION.

In this section we consider briefly the problem of synthesis for discrete systems in which it also takes into account the values of future perturbations (for several steps ahead) obtained with the help of a predicting device.

Consider the extremal problem (1.1)-(1.5) and assume that the controller has been designed and operates in moments t_*, $t_* + h$, ... , $\tau-h$. As a result of the action of the controls

$$u(t_*),\ldots,u(\tau-h)\ ,$$

and the perturbations

$$\omega(t_*,h),\ldots,\omega(\tau-h,h),$$

the system (1.1) at the moment $t=\tau$ is in the state $x(\tau)$. Let it be known that every perturbation $\omega(s,h)$ acting on the system (1.1), after choosing the control $u(s)$, removes the state

$$\check{x}(s+h)= A(s,h)x(s) + b(s,h)u(s)$$

and the system (2.1) enters the new state

$$x(s+h)=\check{x}(s+h)+\omega(s+h)$$

at the moment $s+h$.

Consider that using the accessible information for the moment τ a prediction is made at $\tau+h$, ... , $\tau + \lambda h$ $(\lambda=\lambda(\tau))$; namely that the perturbations are assumed to take the values

$$\check{\omega}(\tau+h|\tau,x(\tau)),\ \ldots\ ,\check{\omega}(\tau + \lambda h|\tau,x(\tau)). \qquad (4.1)$$

Let $u^{o}(t|\tau,x(\tau))$, $t\in T(\tau)$, be the solution to the optimal control problem

$$c'x(t^*) \longrightarrow max, \ x(t+h) = A(t,h)x(t) + b(t,h)u(t),$$

$$x(t)|_{t=\tau} = x(\tau), \qquad h_i'x(t^*) \geq g_i^\tau, \ g_i^\tau = g_i - \tag{4.2}$$

$$- \sum_{t=\tau}^{\tau+\lambda(\tau-h)} h_i'F_h(t^*,t)\check{\omega}(t|\tau-h,x(\tau-h)), i=\overline{1,m},$$

$$u_*(t) \leq u(t) \leq u^*(t), \ t\in T(\tau).$$

We call the construction of the control $u^0(\tau+h|\tau,x(\tau))$ at $\{\tau+h,x(\tau+h)\}$ using $u^0(t|\tau,x(\tau))$, $t\in T(\tau)$, $\omega(\tau,h)$ and the prediction (4.1) the synthesis at $\tau+h$ of the optimal controller with perturbation prediction.

The solution to the synthesis problem is based on the program solution of (4.2) (see Sections 3.1 and 3.3).

Without restriction of generality we may consider that, concerning the initial state $x(\tau)$ for the trajectory $x^0(t)$, $t\in T(\tau)$, generated by control $u^0(t|\tau,x(\tau))$, $t\in T(\tau)$, and the values

$$\check{\omega}(\tau), \ \ldots \ ,\check{\omega}(\tau+\lambda(\tau-h)h),$$

the equalities

$$h_i'x^0(t^*) = g_i, i \in I_{sup}(\tau) \ (\ v(i) \leq 0, \ i \in I_{sup}(\tau))$$

hold.

The set of values of the optimal control at the support moments $u^0_{sup}=(u^0(t|\tau,x(\tau)),t\in T_{sup})$ is calculated as usual,

$$u^0_{sup}=Q(\tau)g_{sup}(\tau),$$

where

$$g_{sup}(\tau) = \begin{bmatrix} g_i(\tau) \\ i \in I_{sup}(\tau) \end{bmatrix},$$

$$g_i(\tau) = g_i^{\tau} - h_i' F_h(t^*, \tau-h)x(\tau) -$$

$$- \sum_{t \in T_N(\tau)} h_i' F_h(t^*, t) b(t,h) u^o(t|\tau, x(\tau)).$$

We have the following information about the optimal support control $\{ u^o(\ |\tau,x(\tau)),\ S_{sup}(\tau)\}$:

$$T_{N+}(\tau) = \{ t \in T_N(\tau) : \Delta(t) > 0,\ \Delta(t)\Delta(t-h) < 0 \} \cup$$

$$\cup \{ t \in T_N(\tau) : \Delta(t) > 0,\ t-h \in T_{sup}(\tau)\},$$

$$T_{N-}(\tau) = \{ t \in T_N(\tau) : \Delta(t) < 0,\ \Delta(t)\Delta(t-h) < 0 \} \cup$$

$$\cup \{ t \in T_N(\tau) : \Delta(t) < 0,\ t-h \in T_{sup}(\tau)\}.$$

$$F_h(t^*, t),\ t \in T_{sup}(\tau) \cup \tau_N \cup (t^*-h) \cup (\cup_{i=0}^{\lambda(\tau)}(\tau+ih)).$$

We can take the array

$$C^k(\tau+h)=\{u^{(k)}(t), t\in T(\tau+h);\ W^k; S^k_{sup} = \{I^k_{sup}, T^k_{sup}\}; T^k_{N+}; T^k_{N-};$$

$$\Phi^k(t),\ t \in T^k_{sup}(\tau) \cup \tau_N \cup (t^*-h) \cup (\cup_{i=0}^{\lambda(\tau)}(\tau+ih));$$

$$\Delta g^k;\ \psi^k(t), t\in T^k_{N+} \cup T^k_{N-} \cup (t^*-h);\ \upsilon^k;\ Q^k\}$$

as the state of the algorithm on k-th iteration at $\tau+h$. As an initial state $C^o(\tau+h)$ at the moment $\tau+h$ the array with the following components can be chosen :

$$u^{(o)}(t)=u(t|\tau,x(\tau)), t\in T(\tau+h);\quad W^o=Hx^o(t^*)-g^{\tau};$$

$$H = H(I,J) = \begin{bmatrix} h'_i(J) \\ i\in I \end{bmatrix}, \quad g^\tau = \begin{bmatrix} g_i^\tau \\ i\in I) \end{bmatrix},$$

$$S^0_{sup} = S_{sup}(\tau) = \{I_{sup}(\tau), T_{sup}(\tau)\};$$

$$T^0_{N+} = T_{N+}(\tau); T^0_{N-} = T_{N-}(\tau); \quad \Phi^0(t) = F_h(t^*, t),$$

$$t \in T^0_{sup}(\tau) \cup \tau_N \cup (t^* - h) \cup (\cup_{i=0}^{\lambda(\tau)} (\tau + ih));$$

$$\Delta g^0 = -H\,[\sum_{t=\tau+h}^{\tau+\lambda(\tau)h-1} \Phi^0(t)(\overset{\vee}{\omega}(t|\tau, x(\tau)) - \overset{\vee}{\omega}(t|\tau-h, x(\tau-h)))+$$

$$+ \quad \Phi^0(\tau)(\overset{\vee}{\omega}(\tau|\tau-h, x(\tau-h)) - \omega(\tau, h)) +$$

$$+ \Phi^0(\tau+\lambda(\tau)h)\ \overset{\vee}{\omega}(\tau+\lambda(\tau)h)|\tau, x(\tau)];$$

$$\psi^0(t) = \psi(t|\tau), t \in T_{N+}(\tau) \cup T_{N-}(\tau) \cup (t^* - h);$$

$$\upsilon^0 = \upsilon(\tau); \quad Q^0 = Q(\tau).$$

It is clear that formulation of the algorithm using the predicting device is not difficult (see Section 3.1, 3.3), so we omit its description.

Now we pass to an example from which one can see the usefulness of predicting devices in some situations.

Example 4.1. Consider the discrete analogue of the problem on the acceleration of the material point

$$x_2(3) \longrightarrow max, \quad x_1(t+h) = x_1(t) + hx_2(t),$$

$$x_2(t+h) = x_2(t) + hu(t), x_1(3) \le 2, \quad x_1(0) = x_2(0) = 0, \qquad (4.3)$$

$$0 \le u(t) \le 1, \quad t = 0, h, \ldots, 2.75, \quad h = 0.25.$$

We shall assume that in the process of control the perturbations

$$\omega_1(t,h) = -t^2/4 + t/2,\ t \in T_1 = \{0, 0.25, \dots, 2\},$$

$$\omega_1(t,h) = 0,\ t \in T_2 = \{2.25, \dots, 3\},\ \omega_2(t,h) = 0,\ t \in T,$$

act on the system.

The prediction on the segment T_1 is constructed on the basis of parabolic extrapolation of the curve $\omega_1(t,h)$ and the value of perturbation at two previous moments of quantization is used ($\lambda(t) = 1,\ t \in T_1$).

The initial control $u_1^0(\cdot)$ (Fig. 4.1.) was chosen.

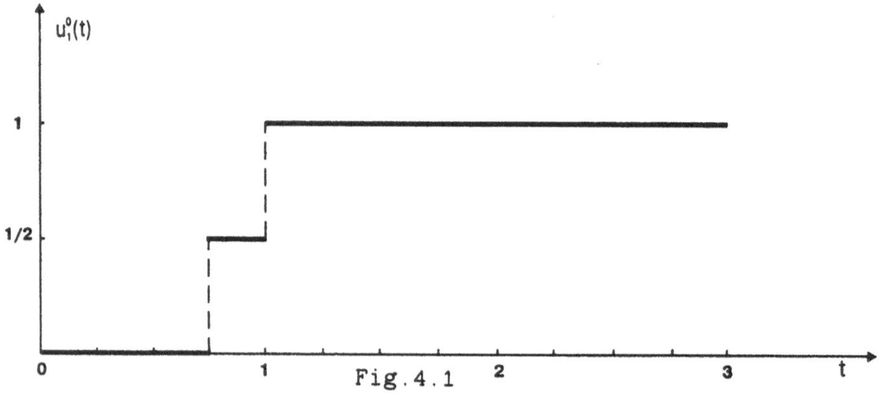

Fig.4.1

The optimal controller constructs the control $u_2^0(\cdot) = (0,0,0,0,37/56,11/16,7/10,25/32,1,1,1,1)$ (Fig. 4.2). Its efficiency is equal to $J(u_2^0(\cdot)) \approx 1.707$.

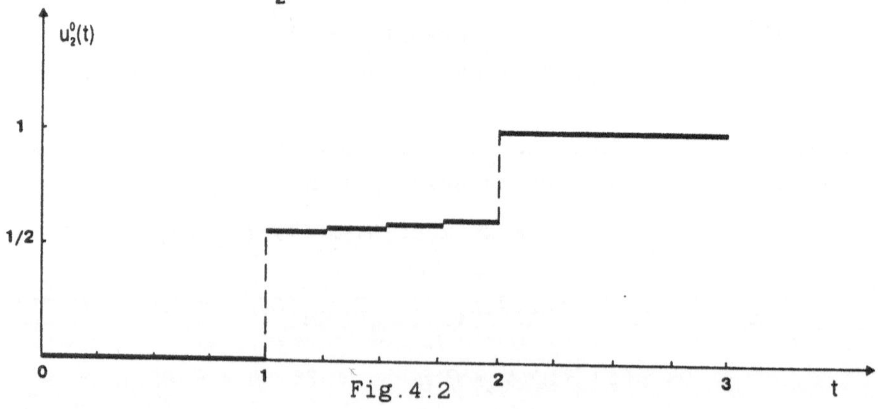

Fig.4.2

If we do not use the procedure of prediction ($\lambda(t) = 0$, $t \in T$) then the optimal control $u_3^0(\cdot) = (0, 0, 0, 3/64, 41/56, 1/4, 5/16, 5/8, 17/24, 1, 1, 1)$ has the form shown in Fig. 4.3 ($J(u_3^0(\cdot)) \approx 1.459$).

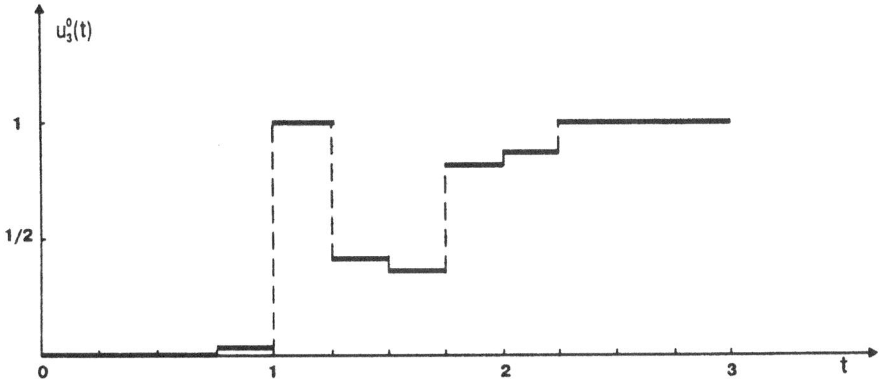

Fig.4.3

If the information on perturbations was known before the beginning of the control process then the optimal control would be $u_4^0(\cdot)=(0,0,0,0,1/14,1,1,1,1,1,1,1)(J(u_4^0(\cdot)) \approx 1.768)$.

In this example an additional utilization of the prediction allows us to increase the efficiency of control for $J(u_2^0(\cdot))-J(u_3^0(\cdot)) \approx 0.248$.

3.5. STABILITY, STABILIZATION, OPTIMALITY.

Stability of solutions [35] is one of the central problems of the theory of differential equations which have been formulated in numerous applications. For a long time the results obtained in this field have been widely used in the theory of control for stabilization of unstable dynamic systems [3]. In the classical era of the automatic regulation theory one was often satisfied with the achievement of the common property of stability for the systems. With the development of technology the requirements for stabilization have been increasing. One has begun to evaluate stable transitional processes on different quality criteria. The new stage of the stabilization theory came after construction of the optimal control theory [38]. The first large-scale and practically important application of the theory of optimal processes to stabilization problems was the Lyotov-Kalman method of analytical construction of optimal regulators [24,34]. It is effectively used for stabilization of linear systems with the square estimate of quality of transitional processes, without taking into account any constraints on controls and trajectories. The use of other results of the mathematical theory of optimal processes for the solution of complicated stabilization problems is becoming restricted because of the absence of effective algorithms for synthesis of optimal systems.

In this section we shall show how results for optimal controllers can be applied to the construction of optimal stabilizers with restricted control.

Consider an n-dimensional discrete process $x(t)$, $t \in \{ 0, h, \ldots, \}$ which is described by

$$x(t+h) = A(h)x(t) \quad . \tag{5.1}$$

We shall suppose that the system (5.1) is unstable relative to the stationary state $x = 0$, i.e. there is an (unstable) initial state $x_0 \neq 0$, such that the process $x(t)$, $t \geq 0$, started at the moment $t = 0$ from the state x_0 has the property

$$||x(t)|| \longrightarrow \infty \text{ if } t \longrightarrow \infty .$$

Let for stabilization of system (5.1) a scalar bounded control $u(t)$, $t \geq 0$, be admissible:

$$|u(t)| \leq 1, \; t \geq 0 . \tag{5.2}$$

We consider that the interaction between the control (5.2) and the object of stabilization (5.1) is described as

$$x(t + h) = A(h)x(t) + b(h)u(t) \tag{5.3}$$

Denote

$$X_\alpha = \{ x \in R^n : |x_j| \leq \alpha, \; j = \overline{1,n} \} , \; (0 < \alpha < \infty). \tag{5.4}$$

Definition 5.1. Let us call the state $x(0) \in X_\alpha$ stabilized if there exists a control $u(t)$, $t \in T_{\theta-h} = \{ 0, h, \ldots, \theta - h\}$ which satisfies the constraint (5.2) and generates the trajectory $x(t)$, $t \in T_\theta$, along which the condition

$$x(\theta) \in X_\alpha \tag{5.5}$$

is fulfilled.

Let us call the system (5.1) $\alpha\theta$-stabilized if all its states $x(0) \in X_\alpha$ are $\alpha\theta$-stabilized.

If in addition to (5.5) the relation $x(t) \in X_{\rho\alpha}$, $t \in T$, $(\rho \geq 1)$ is fulfilled, we speak about the uniform $\alpha\theta$-stabilizability.

Definition 5.2. Let us call the state $x(0) \in X_\alpha$ $\alpha q\theta$ - asymptotically stabilized if there exists a control $u(t)$, $t \in T_{\theta-h}$, which satisfies constraint (5.2) and generates the trajectory $x(t)$, $t \in T_\theta$, along which the condition

$$x(\theta) \in X_{q\alpha}$$

is fulfilled.

We shall call the system (5.1) $\alpha q\theta$-asymptotically stabilized if all its states $x(0)\in X_{\alpha}$ are $\alpha q\theta$-asymptotically stabilized $(0 < q < 1)$.

In terms of the optimal control theory we have considered the problems of existence of stabilizing controls. For the construction of the latter it would be possible to introduce the first phase of the method in the spirit of linear programming. However, let us pass immediately to optimization.

Definition 5.3. For the given initial state $x(0)\in X_{\alpha}$ the $\alpha q\theta$-asymptotically stabilizing control will be called optimal if the value of the parameter q is minimal on it, i.e. if it is a solution to the problem

$$q \longrightarrow min, \quad x(t+h) = A(h)x(t) + b(h)u(t),$$

$$x(0) = x_0, \quad |x_j(\theta)| \le q\alpha, \ j = \overline{1, n}, \tag{5.6}$$

$$|u(t)| \le 1, \ t \in T_{\theta-h}.$$

The optimal uniformly $q\alpha\theta$-asymptotically stabilizing control is a solution to the problem

$$q \longrightarrow min, \quad x(t+h) = A(h)x(t) + b(h)u(t),$$

$$x(0) = x, \ x(t) \in X_{\rho\alpha}, \ t \in T_{\theta}; \ |x_j(\theta)| \le q\alpha, \ j = \overline{1, n}, \tag{5.7}$$

$$|u(t)| \le 1, \ t \ge T$$

The optimal controls introduced are program controls.

We describe the scheme of synthesis of stabilizing controls which are generated in real time.

The control system (5.2),(5.3) is assumed to operate at the moments $0,h,\dots,\tau$. Being under the action of controls $u(0), \dots ,u(\tau-h)$ and external perturbations $\omega(0), \dots, \omega(\tau-h)$ it is in the state $x(\tau)$ at the moment τ. At the next moment $\tau+h$ it will appear not in the state

$$\check{x}(\tau+h) = A(h)x(\tau) + b(h)u(\tau)$$

but in

$$x(\tau+h) = \check{x}(\tau+h) + \omega(\tau).$$

The device which for each $\tau \geq 0$ and any $x(\tau+h)$ solves the extremal problem

$$q \longrightarrow min, \quad x(t+h) = A(h)x(t) + b(h)u(t),$$

$$|x_j(\tau+\theta+h)| \leq q\alpha, \ j = \overline{1, n}, \ x(\tau+h) = \check{x}(\tau+h) + \omega(\tau), \qquad (5.8)$$

$$|u(t)| \leq 1, \ t \in \{\tau+h, \ \ldots \ , \tau+\theta\},$$

will be called the optimal stabilizer.

The definition of the uniform optimal stabilizer can be introduced by analogy.

It is evident that the direct solution of problem (5.7) at each moment τ is too tedious and cannot be really performed at the accepted time h. Therefore the optimal control $u^o(\cdot|\tau+h,x(\tau+h))$ of problem (5.7) will be constructed with the help of correction of the control $u^o(\cdot|\tau,x(\tau))$. As a basis for the stabilizer at the initial moment $\tau = 0$ it is possible to choose the optimal program $u^o(\cdot)$ for (5.6).

According to [19] the solution to problem (5.7) having been constructed for the moment τ is the set

$$\{ u^o(\cdot|\tau,x(\tau)), S_{sup}(\tau)\}$$

where

$$S_{sup}(\tau) = \{ J_{sup}(\tau), T_{sup}(\tau)\},$$

$$J_{sup}(\tau) \subset J = \{ 1,2,\ldots,n\},$$

$$T_{sup}(\tau) = \{\tau_1, \ \ldots \ ,\tau_l\},$$

$$\tau \leq \tau_1(\tau) < \tau_2(\tau) < \ \ldots \ < \tau_l(\tau) \leq \tau + \theta.$$

In addition, the correlations

$$|J_{sup}(\tau)| = |T_{sup}(\tau)| = l,$$

$$0 \le l \le n, \ det\ P(\tau) \ne 0$$

are fulfilled. Here $P(\tau)$ is the $l \times l$ matrix constructed by the system elements (5.1).

Correction of the solution $\{u^0(\cdot|\tau,x(\tau)),\ S_{sup}(\tau)\}$ consists of recounting $\tau_1, \ldots, \tau_l$. These numbers in the general case differ a little from the corresponding moments $\tau_1(\tau), \ldots, \tau_l(\tau)$ which are known for the stabilizer at τ. The algorithm can be elaborated on the basis of results obtained in Sections 3.1-3.4.

CHAPTER 4

CONSTRUCTING OPTIMAL FEEDBACK CONTROLS

In this chapter we generalize the approach proposed in previous chapters. We shall consider the synthesis problem under conditions which are closed maximally to real processes. We shall construct optimal feedback controls assuming that we have only incomplete and inexact measurements of output signals.

4.1. SYNTHESIS OF OPTIMAL CONTROLS ON INEXACT MEASUREMENTS OF OUTPUT SIGNALS.

We consider a discrete linear system the behaviour of which on the discrete interval $T(t_*)= \{t_*, t_*+h, \ldots, t^*-h\}$ described by

$$x(t+h)=A(t,h)x(t) + b(t,h)u(t). \qquad (1.1)$$

The initial state of system (1.1) is supposed to be known inexactly. A priori information about it has the form

$$x(t_*) = z \in \check{X}_* = \{ z \in \mathrm{R}^n : Gz = f,\ d_* \le z \le d^* \}, \qquad (1.2)$$

$$(f \in \mathrm{R}^r,\ rank\, G = r \le n).$$

The family of trajectories of system (1.1)

$$\check{X}(t|u(\cdot))= \{ x(t|\ z,u(\cdot)), z \in \check{X}_* \},\ t\in T^*(t_*)=T(t_*) \cup t^* .$$

corresponds to each control $u(t), t\in T(t_*)$, limited by

constraints

$$u_*(t) \le u(t) \le u(t^*), \ t\in T(t_*). \tag{1.3}$$

Let in state space the terminal set

$$X^* = \{ x\in R^n : h_i'x \ge g_i, i=\overline{1,m} \}. \tag{1.4}$$

be given.

Following the principle of getting the guaranteed result, the control $u(\cdot)=(u(t),t\in T(t_*))$, will be called admissible, if the corresponding movement $\check{X}(t|u(\cdot))$, $t \in T(t_*)$, satisfies the terminal inclusion

$$\check{X}(t^*|u(\cdot)) \subseteq X^*. \tag{1.5}$$

In the spirit of the accepted approach the value of the quality criterion for admissible control $u(\cdot)$ is

$$J(u(\cdot))= \min_{z\in \check{X}_*} h_0'x(t^*|z,u(\cdot)). \tag{1.6}$$

Admissible control $u^0(t),t\in T(t_*)$, having the property

$$J(u^0(\cdot))= \max J(u(\cdot)) \tag{1.7}$$

is called optimal.

Problem (1.1)-(1.7) does not always have a solution due to indefiniteness (1.2) as it is often impossible to satisfy (1.5). On the other hand, if admissible controls exist, the efficiency (1.7) of the optimal control may be low for the same reasons.

To increase the control efficiency the procedure of system optimization is supplemented by the measuring device

$$y(t)= c'(t)x(t)+\xi(t),(y\in R^1). \tag{1.8}$$

Assume the measurement errors $\xi(t)$, $t \in T(t_*)$ satisfy restrictions

$$\underline{\xi}(t) \le \xi(t) \le \overline{\xi}(t), \quad t \in T(t_*). \tag{1.9}$$

The measuring device (1.8),(1.9) is considered to have recorded the signal $y_\tau(\cdot) = (y(t), t=t_*, t_*+h, \dots, \tau)$. Let us verify by it the a priori distribution $\check{X}_*$ of initial states.

The set $\hat{X}_*^\tau = \hat{X}_*(y_\tau(\cdot))$ is called the a posteriori distribution of initial states corresponding to the observation process up to the moment τ if it consists of those and only those initial states $x(t_*) \in \check{X}_*$ which can generate the observed signal $y_\tau(\cdot)$, together with some measurement errors $\xi(t), t \ge t_*$, and control $u(\cdot)$.

In itself set $\hat{X}_*^\tau$ is not necessary for solving the synthesis problem. We shall need only its numerical characteristics (estimates) connected with the terminal states:

$$\hat{\alpha}^\tau(t^*) = \hat{\alpha}^\tau(t^* | u(\cdot)) = \min_{z \in \hat{X}_*^\tau} h_i' x(t^* | z, u(\cdot)), \quad i=\overline{0,m}. \tag{1.10}$$

Calculation of estimates $\hat{\alpha}_i^\tau(t^*), i=\overline{0,m}$, will be called the τ-observation problem accompanying original problem (1.1)-(1.7).

Control $\hat{u}(\cdot) = (\hat{u}(t), t \in T(t_*))$, with the known starting part $u(t)$, $t_* \le t \le \tau-h$, is called τ-a posteriori admissible if

$$\hat{\alpha}_i^\tau(t^*) \ge g_i, \quad i=\overline{1,m}. \tag{1.11}$$

Define τ-a posteriori optimal control $\hat{u}^o(\cdot)$ by

$$\hat{\alpha}_o^\tau(t^* | \hat{u}^o(\cdot)) = \max_{\hat{u}(\cdot)} \hat{\alpha}_o^\tau(t^* | \hat{u}(\cdot)). \tag{1.12}$$

The search of controls $\hat{u}^o(\cdot), t=\tau, \tau+h, \dots, t^*-h$, will be called the τ-problem of optimal control accompanying problem (1.1)-(1.7).

As a whole, problem (1.1)-(1.12) is called the problem of optimal control on incomplete and inexact measurements of system states.

In this chapter solutions of two types are given: program solution for any fixed $\tau \in T(t_*)$ and optimal feedback solution consisting of optimal estimator (see Section 2.4) and optimal controller (see Chapter 3).

4.2. PROGRAM SOLUTION OF τ-OBSERVATION PROBLEM.

Except for the mathematical model (1.1)-(1.7) and the control $u(t)$ used on the interval $\{t_*, t_*-h, \dots, \tau-h\}$, let the signal $y(t)$, $t_* \le t \le \tau$, written as (1.8),(1.9), be known.

Denote the fundamental matrix by $F_h(t,\tau), t,\tau \in T(t_*)$. Let $x(t)$, $t_* \le t \le \tau$ be a control system trajectory and

$$x_u(t+h)=A(t,h)x_u(t) + b(t,h)u(t), x(t_*)=0,$$

$$y_0=(t)=y(t)-c'x_u(t), t_* \le t \le \tau.$$

Since

$$\hat{\alpha}_i^\tau(t^*)= \hat{\alpha}_i^\tau(t^* | u(\cdot)) = \min_{z \in \hat{X}_*^\tau} h_i' F(t^*, t_*-h)z + h_i' x_u(t^*)$$

the problem of τ-observation (1.10) is reduced to the extremal problems

$$\hat{\gamma}_i^\tau(t^*) = \min_z h' F(t^*, t_*-h)z,$$

$$\xi_*(t) \le y_0(t) - c'F(t,t_*-h)z \le \xi^*(t), \quad t_* \le t \le \tau, \tag{2.1}$$

$$Gz = f, \; d_* \le z \le d^*, \; i = \overline{0,m}.$$

At the same time

$$\hat{\alpha}_i^\tau(t^*) = h_i' x_u(t^*) + \hat{\gamma}_i^\tau(t^*), \; i = \overline{0,m}.$$

Denote

$$a'(t) = (a_1(t), a_2(t), \dots, a_n(t))' = -c'(t)F(t, t_*-h),$$

$$\eta_i = -h_i' F(t^*, t_*-h), \; i = \overline{0,m}.$$

Then problem (2.4) can be written in the form

$$\hat{\gamma}_i^{\tau} = \max_z \eta_i' z,$$

$$\xi_*(t) \le y_0(t) + a'(t)z \le \xi^*(t), \; t_* \le t \le \tau, \tag{2.2}$$

$$Gz = f, \; d_* \le z \le d^*, \; i = \overline{0,m}.$$

By virtue of uniformity of problem (2.2), index i will be omitted in future and we shall consider an arbitrary problem of family (2.2).

Solve problem (2.2) by linear programming methods (see the Appendix). Let $\{z(\tau), S_{sup}(\tau)\}$ be an optimal feasible solution of problem (2.2). The optimal support $S_{sup}(\tau) = \{J_{sup}(\tau), T_{sup}(\tau)\}$ is a totality from the set $J_{sup}(\tau) \subset J = \{1, 2, \ldots, n\}$ of supporting indices of the feasible solution $z(\tau)$ and the set $T_{sup}(\tau) \subset T^{\tau} = \{ t : t_* \le t \le \tau \}$ of supporting moments $t_* \le \theta_1 = \theta_1(\tau) < \ldots < \theta_l = \theta_l(\tau) \le \tau$. Relations

$$r + |T_{sup}(\tau)| = |J_{sup}(\tau)|, \; 0 \le l \le n-r, \quad \det P \ne 0,$$

$$P = P(\tau) = P(\{T_{sup}(\tau), M\}, J_{sup}(\tau)) = \begin{bmatrix} a_j(t) : j \in J_{sup}(\tau) \\ t \in T_{sup}(\tau) \\ G(M, J_{sup}(\tau)) \end{bmatrix}$$

are carried out.

Introduce designations

$$Q = Q(\tau) = Q(J_{sup}(\tau), \{T_{sup}(\tau), M\}) = P^{-1}(\tau) =$$

$$= \begin{pmatrix} ((q_j(t) : t \in T_{sup}(\tau), (q_{ji} : i \in M))' \\ j \in J_{sup}(\tau) \end{pmatrix}.$$

Construct sets

$$T_N = T_N(\tau) = T^{\tau} \setminus T_{sup}(\tau); \quad J_N = J_N(\tau) = J \setminus J_{sup}(\tau).$$

To every moment $t \in T^{\tau}$ and the indices $j \in J$, $i \in M$, we add the numbers

$$\nu(t) = \nu(t|\tau), \; \Delta_j = \Delta_j(\tau), \; \mu_i = \mu_i(\tau) :$$

$$\nu(t) = 0,\ t \in T_N(\tau);\ \Delta_j(\tau) = 0,\ j \in J_{sup}(\tau);$$

$$\mu = \mu(\tau) = (\mu_i(\tau),\ i \in M;$$

$$\nu_{sup} = (\nu(T_{sup}(\tau)) = (\nu(\theta_1(\tau)),\ \nu(\theta_2(\tau)),\ \ldots,\ \nu(\theta_l(\tau)));$$

$$\nu_N = \nu(T_N(\tau)) = (\nu(t),\ t \in T_N(\tau));\ \lambda_{sup} =$$

$$= (\nu(T_{sup}(\tau)),\ \mu(\tau)) = (\nu_{sup},\ \mu);$$

$$\lambda_{sup} = \eta'_{sup} Q(\tau),\ \eta_{sup} = (\eta_j,\ j \in J_{sup}(\tau));\ \lambda_N = \lambda(T_N(\tau)) =$$

$$= (\nu(T_N(\tau)),\ \mu(\tau));\ \lambda = \lambda(T^\tau) = (\lambda(t),\ t \in T^\tau) =$$

$$= (\lambda(T_{sup}(\tau)),\ \lambda(T_N(\tau))) = (\lambda_{sup},\ \lambda_N);$$

$$\Delta'_{sup}(J_N) = \Delta'(\tau | J_N(\tau)) = (\Delta_j(\tau),\ j \in J_N(\tau))' =$$

$$= \nu'_{sup} A(T_{sup}(\tau),\ J_N(\tau)) + \mu' G(M,\ J_N(\tau)) - \eta'_N,$$

$$\eta_N = \eta_N(\tau) = (\eta_j,\ j \in J_N(\tau)).$$

A feasible solution $z(\tau)$ is optimal iff there exists a support $S_{sup}(\tau)$ such that

$$\Delta_j(\tau) \le 0 \text{ if } z_j(\tau) = d^*_j;\quad \Delta_j(\tau) \ge 0 \text{ if } z_j(\tau) = d_{*j};$$

$$\Delta_j = 0 \text{ if } d_{*j} \le z_j(\tau) \le d^*_j,\quad j \in J_N(\tau);$$

$$\nu(\theta_k(\tau)) \ge 0 \text{ if } y_0(\theta_k(\tau)) + a'(\theta_k(\tau))z(\tau) = \xi^*(\theta_k(\tau));$$

$$\nu(\theta_k(\tau)) \le 0 \text{ if } y_0(\theta_k(\tau)) + a'(\theta_k(\tau))z(\tau) = \xi_*(\theta_k(\tau));$$

$$\nu(\theta_k(\tau)) = 0 \text{ if } \xi_*(\theta_k(\tau)) \le y_0(\theta_k(\tau)) + a'(\theta_k(\tau))z(\tau) \le \xi^*(\theta_k(\tau)),$$

$$k = \overline{1,\ l}.$$

4.3. SYNTHESIS OF OPTIMAL ESTIMATOR.

Assume that the problem

$$\eta'z \longrightarrow \max\ ,\ Gz = f\ ,\ d_* \le z \le d^*,$$
$$\xi_*(t) \le y_0(t) + a'(t)z \le \xi^*(t),\ t_* \le t \le \tau - h, \tag{3.1}$$

has been solved using output signals $y(t)$, $t_* \le \le t \le \tau-h$ and also the values of controlling influence $u(t), t_* \le t \le \tau-2h$, produced by the controller (see below) and that $\{z(\tau-h), S_{sup}(\tau-h)\}$ is an optimal solution to this problem.

We give the estimates found from (3.1) (at $\eta = \eta_i, i=\overline{0,m}$) to controller which will produce $u(\tau-h)$ for $\tau-h$. Write the signal $y(\tau)$ of measuring device (1.8),(1.9) at the moment τ. Proceeding from this information we find optimal solution $\{z(\tau), S_{sup}(\tau)\}$ to the problem

$$\eta'z \longrightarrow \max\ ,\ Gz = f\ ,\ d_* \le z \le d^*,$$
$$\xi_*(t) \le y_0(t) + a'(t)z \le \xi^*(t),\ t_* \le t \le \tau \tag{3.2}$$

where

$$y_0(\tau) = y(\tau) - c'(\tau)x_u(\tau);$$

$$x_u(\tau) = A(\tau-h,h)x_u(\tau-h) + b(\tau-h,h)u(\tau-h).$$

Construction of the optimal solution $\{z(\tau), S_{sup}(\tau)\}$ to problem (3.2) for any $y(\tau)$, proceeding from the optimal solution $\{z(\tau-h), S_{sup}(\tau-h)\}$ of problem (3.1), will be called optimal estimator synthesis at the moment τ.

Now let us begin to solve this problem. According to information available to the moment $\tau - h$ we calculate

$$\omega(\tau - h) = y_0(\tau) + a'(\tau)z(\tau-h). \tag{3.3}$$

If

$$\xi_*(\tau) \le \omega(\tau-h) \le \xi^*(\tau)$$

then

$$\{z(\tau),\ S_{sup}(\tau)\} = \{z(\tau-h), S_{sup}(\tau)\}.$$

Therefore the optimal estimator synthesis problem at the moment τ does not occur, or in another words, it is solved trivially. It occurs at $\omega(\tau-h) \notin [\xi_*(\tau),\xi^*(\tau)]$. Let $w(\tau-h) > \xi^*(\tau)$ (for definiteness) .

Embed problem (3.2) in the family of extremal problems depending on a parameter ρ,

$$\eta'z \longrightarrow max \ , \ Gz = f \ , \ d_* \le z \le d^*,$$

$$\xi_*(t) \le y_0(t) + a'(t)z \le \xi^*(t), \quad t_* \le t \le \tau - h, \tag{3.4}$$

$$\xi_*(\tau) \le y_0(\tau) + a'(\tau)z \le \rho.$$

Problem (3.4) at $\rho = \omega(\tau-h)$ has the solution $\{z(\tau-h), S_{sup}(\tau-h)\}$. To find $\{z(\tau),S_{sup}(\tau)\}$ we shall iteratively decrease the parameter ρ: $\omega(\tau-h) = \rho_0 > \rho_1 > \dots > \rho_p = \xi^*(\tau)$, constructing simultaneously the solutions $\{z^k, S^k_{sup}\} =$

$= \{z(\tau-h)|\rho_k), S_{sup}(\tau-h|\rho_k)\}$ of problem (3.4). Then we set

$\{z(\tau), S_{sup}(\tau)\} = \{z_p, S^p_{sup}\}$.

Proceeding to the description of the optimal estimator we denote by T^k_{sup}, J^k_{sup} the sets of supporting time moments and indices from J on the k-th iteration of the algorithm and let

$$T^k_N = [(\{t_*\} \cup \{t_*+h\} \cup \{t_* \pm h, t \in T^k_{sup}\}) \cap T^\tau],$$

$$L^k_{sup} = \{ T^k_{sup}, M \}.$$

We call the array

$$C^k(\tau-h) = \{z^k; S^k_{sup}; T^k_N; y(T^k_N); u(T^k_{sup}); x_u(T^k_{sup});$$

$$F(T^k_{sup}, t_*-h); Q^k = Q(J^k_{sup}, L^k_{sup});$$

$$\lambda^k = (\nu^k(T^k_{sup}), \mu^k(M)); \Delta^k(J^k_N); \rho_k\}$$

as a state of the algorithm on k-th iteration at the moment $\tau-h$.

Compose from

$$z^0 = z(\tau-h);\ S^0_{sup} = S_{sup}(\tau-h);\ T^0_N = [(\{t_*\} \cup \{t_*+h\} \cup$$

$$\cup \{t\pm h,\ t\in T_{sup}(\tau-h)\}) \cap T^\tau];\ y(T^0_N);\ u(T_{sup}(\tau-h);$$

$$x_u(T_{sup}(\tau-h);\ F(T_{sup}(\tau-h),\ t_*-h);\ Q^0 = Q(\tau-h);$$

$$\lambda^0 = \lambda(\tau-h);\ \Delta^0(J^0_N) = \Delta(\tau-h|J_N(\tau-h));\ \rho_0 = \omega(\tau-h),$$

the zeroth state of the algorithm.

Iteration of the algorithm $C^k(\tau-h) \longrightarrow C^{k+1}(\tau-h)$ consists of the following steps.

Step 1. Verify the condition $\tau\in T^k_{sup}$ If it is fulfilled we proceed to *Step 2*. Otherwise we pass to *Step 5*.

Step 2. Let $q^k(\tau) = Q^k(J^k_{sup},\ \tau) = q^k(\theta_1(\tau-h)) = q^k_1 =$

$$= (q_{j1}, j\in J^k_{sup}).$$

Calculate

$$\beta^k_j = \begin{cases} (z^k_j - d^*_j)/q^k_{j1} & \text{at } q^k_{j1} < 0, \\ (z^k_j - d_{*j})/q^k_{j1} & \text{at } q^k_{j1} > 0, \\ \infty & \text{at } q^k_{j1} = 0,\ j\in J^k_{sup}. \end{cases} \tag{3.5}$$

Set

$$x_u(t) = \begin{cases} A(t-h,h)x_u(t-h) + b(t-h,h)u(t-h), \text{ when } t-h = \theta\in T^k_{sup}, \\ A^{-1}(t,h)x_u(t+h) - b(t,h)u(t), \text{ when } t+h = \theta\in T^k_{sup}, t\in T^k_N. \end{cases}$$

$$x_u(t_*) = 0\ ;\ x_u(t_*+h) = b(t_*,h)u(t_*),$$

$$y_0(t) = y(t) - c'(t)x_u(t),\ t\in T^k_N\ , \tag{3.6}$$

$$a'(t) = \begin{cases} -c'(t)A(t-h,h)F(t-h,t_*-h), \text{ when } t-h = \theta\in T^k_{sup}, \\ -c'(t)A^{-1}(t,h)F(t+h,t_*-h), \text{ when } t+h = \theta\in T^k_{sup},\ t\in T^k_N. \end{cases}$$

Construct

$$\beta^k(t^o)=\begin{cases}[y_o(t)+a'(t)z^k-\xi^*(t)]/a'(t)q_1^k, & \text{when } a'(t)q_1^k<0;\\ [y_o(t)+a'(t)z^k-\xi_*(t)]/a'(t)q_1^k, & \text{when } a'(t)q_1^k>0;\\ \infty, \text{ when } a'(t)q_1^k=0, & t\in T_N^k,\end{cases} \tag{3.7}$$

$$\beta^k(\tau)=\rho_k-\xi^*(\tau), \tag{3.8}$$

$$\beta_{j_o}^k=\min \beta_j^k,\ j\in J_{sup}^k;\quad \beta^k(t^o)=\min \beta^k(t),\ t\in T_N^k;$$

$$\beta_o^k=\min\{\beta_{j_o}^k,\ \beta^k(t^o),\ \beta^k(\tau)\}. \tag{3.9}$$

Let

$$z_{sup}^{k+1}=(z_j^{k+1},\ j\in J_{sup}^k)=z_{sup}^k-\beta_o^k\, q_{sup\ 1}^k,$$

$$\rho_{k+1}=\rho_k-\beta_o^k$$

Here

$$q_{sup\ 1}^k=(q_{j1}^k,\ j\in J_{sup}^k);\quad z_{sup}^k=(z_j^k,\ j\in J_{sup}^k);$$

$$z^{k+1}=(z_{sup}^{k+1},\ z_N^k);\quad z_N^k=(z_j^k,\ j\in J_N^k).$$

The following cases are possible

a) $\beta_o^k=\beta_{j_o}^k$; b) $\beta_o^k=\beta^k(t^o)$; c) $\beta_o^k=\beta^k(\tau)$.

If Case a) is realized we go to *Step 3*. In Case b) we proceed to *Step 4*. In Case c) we pass to correlation (3.27) of *Step 6*.

Step 3. Calculate

$$\Delta\lambda^{k'}=(\Delta\nu^k,\ \Delta\mu^k)'=(\Delta\nu^k(T_{sup}^k),\ (\Delta\mu_j^k,\ j\in M))'=$$

$$=e'_{j_o}Q^k(J_{sup}^k,\ L_{sup}^k)\ \mathrm{sign}\ q_{j_o}^k,$$

$$e_{j_o}=(e_j:\ e_j=0,\ j\neq j_o,\ e_{j_o}=1,\ j\in J_{sup}^k), \tag{3.10}$$

$$\Delta\delta^{k'}=\Delta\delta^{k'}(J)=\Delta\lambda^{k'}P^k(L_{sup}^k,\ J);$$

$$\sigma^k(t) = \begin{cases} -\nu^k(t)/\Delta\nu^k(t) & \text{when } \nu^k(t)\Delta\nu^k(t) < 0 , \\ \infty & \text{when } \nu^k(t)\Delta\nu^k(t) \geq 0 , \ t \in T^k_{sup} ; \end{cases}$$

$$\sigma^k_j = \begin{cases} -\Delta^k_j/\Delta\delta^k_j & \text{when } \Delta^k_j\Delta\delta^k_j < 0, \\ \infty & \text{when } \Delta^k_j\Delta\delta^k_j \geq 0 , \ j \in J^k_N ; \end{cases} \tag{3.11}$$

$$\sigma^k(t^0) = min \ \sigma^k(t) \ t \in T^k_{sup} ; \quad \sigma^k_{j_*} = min \ \sigma^k_j , \ j \in J^k_N ;$$

$$\sigma^k_0 = min \{ \sigma^k(t^0), \sigma^k_j \} \tag{3.12}$$

Let

$$S^{k+1}_{sup} = \{ T^{k+1}_{sup} , J^{k+1}_{sup} \} ; \quad T^{k+1}_{sup} = T^k_{sup} \backslash t^0 ; \ J^{k+1}_{sup} = J^k_{sup} \backslash j_0 ,$$

$$\text{when } \sigma^k_0 = \sigma^k(t^0), \tag{3.13}$$

$$T^{k+1}_{sup} = T^k_{sup} ; \quad J^{k+1}_{sup} = (J^k_{sup} \backslash j_0) \cup j_* , \quad \text{when } \sigma^k_0 = \sigma^k_{j_*} ,$$

$$L^{k+1}_{sup} = \{ T^{k+1}_{sup}, M \} \tag{3.14}$$

Let situation (3.13) be represented. Then

$$Q^{k+1} = Q^{k+1}(J^{k+1}_{sup} , L^{k+1}_{sup}) = Q^k(J^k_{sup} \backslash j_0, L^k_{sup} \backslash t^0) -$$

$$- Q^k(J^k_{sup} \backslash j_0, t^0) \ Q(j_0, L^k_{sup} \backslash t^0)/q^k_{j_0 1_0} , \tag{3.15}$$

$$q^k_{j_0 1_0} = Q^k(j_0, j_0)$$

$$\begin{cases} \nu^{k+1}(T^{k+1}_{sup}) = \nu^k(T^k_{sup} \backslash t^0) + \sigma^k_0 \Delta\nu^k(T^k_{sup} \backslash t^0) , \\ \mu^{k+1}(M) = \mu^k(M) + \sigma^k_0 \Delta\mu^k(M) , \end{cases} \tag{3.16}$$

$$\begin{cases} \Delta^{k+1}(J_N^{k+1}\backslash j_0) = \Delta^k(J_N^k) + \sigma_0^k \Delta\delta^k(J_N^k) , \\ \Delta_{j_0}^{k+1} = \sigma_0^k \Delta\delta_{j_0}^k = \sigma_0^k \; sign \; q_{j_0 1}^k \end{cases}$$

If situation (3.14) takes place then

$$Q^{k+1} = Q^{k+1}(J_{sup}^{k+1}, L_{sup}^{k+1}) = Q^k(J_{sup}^k, L_{sup}^k) -$$

$$- Q^k(J_{sup}^k, L_{sup}^{k+1}) [P^k(L_{sup}^k, j_*) - P^k(L_{sup}^k, j_0)]\times$$

$$\times Q^k(j_0, L_{sup}^k)/[Q^k(j_0, L_{sup}^k)P^k(L_{sup}^k, j_*)] ; \quad (3.17)$$

$$\begin{cases} \nu^{k+1}(T_{sup}^{k+1}) = \nu^k(T_{sup}^k) + \sigma_0^k \Delta\nu^k(T_{sup}^k), \\ \mu^{k+1}(M) = \mu^k + \sigma_0^k \Delta\mu^k, \end{cases}$$

$$\begin{cases} \Delta^{k+1}(J_N^{k+1} j_0) = \Delta^k(J_N^k\backslash j_*) + \sigma_0^k \Delta\delta^k(J_N^k\backslash j_*) , \\ \Delta_{j_0}^{k+1} = \sigma_0^k sign \; q_{j_0 1}^k \end{cases}$$

Proceed to *Step 6*.

Step 4. Having calculated $a(t^0)$, $x_u(t^0)$, according to (3.6) we construct

$$\Delta\lambda^{k'} = (\Delta\nu^k, \Delta\mu^k)' = (\Delta\nu^k(T_{sup}^k), (\Delta\mu_j^k, j\in M))' =$$

$$= a'_{sup}(t^0)Q^k(J_{sup}^k, L_{sup}^k) \; sign \; (a'_{sup}(t^0)q_1^k) , \quad (3.19)$$

$$\Delta\delta^{k'} = \Delta\delta^{k'}(J) = \Delta\lambda^{k'} P^k(L_{sup}^k, J) - a'(t^0) \; sign \; (a'_{sup}(t^0)q_1^k)$$

Following (3.10)-(3.12),(3.19) we find σ_0^k and change the support $S_{sup}^k \longrightarrow S_{sup}^{k+1}$:

$$T_{sup}^{k+1} = (T_{sup}^k\backslash t^*) \cup t^0 \; ; \; J_{sup}^{k+1} = J_{sup}^k \; ; \; \sigma_0^k = \sigma^k(t^*) \quad (3.20)$$

$$T^{k+1}_{sup} = T^{k}_{sup} \cup t^{0};\ J^{k+1}_{sup} = J^{k}_{sup} \cup j_{*}\ ; \qquad \sigma^{k}_{0} = \sigma^{k}_{j_*}\ , \qquad (3.21)$$

$$L^{k+1}_{sup} = \{\ T^{k+1}_{sup},\ M\ \}$$

If the situation (3.20) is realized then having put

$$P^{k}(t^{0},\ J^{k}_{sup}) = (a_{j}(t^{0}),\ j\in J^{k}_{sup})$$

we get

$$Q^{k+1} = Q^{k+1}(J^{k+1}_{sup}\ ,\ L^{k+1}_{sup}\ ) = Q^{k}(J^{k}_{sup}\ ,\ L^{k}_{sup}\ ) -$$

$$- Q^{k}(J^{k}_{sup}\ ,\ t^{*}\ )\ [P^{k}(t^{*},J^{k}_{sup}\ ) - P^{k}(t^{0},\ J^{k}_{sup}\ )]\times \qquad (3.22)$$

$$\times Q^{k}(J^{k}_{sup}\ ,\ L^{k}_{sup}\ )/[-P^{k}(t^{0},\ J^{k}_{sup}\ )Q^{k}(J^{k}_{sup}\ ,\ t^{*})]\ ;$$

$$\begin{cases} \nu^{k+1}(T^{k+1}_{sup}\backslash t^{0}) = \nu^{k}(T^{k}_{sup}\backslash t^{*}) + \sigma^{k}_{0}\Delta\nu^{k}(T^{k}_{sup}\backslash t^{*})\ , \\ \mu^{k+1}(M) = \mu^{k}(M) + \sigma^{k}_{0}\Delta\mu^{k}(M)\ , \end{cases}$$

$$(3.23)$$

$$\begin{cases} \Delta^{k+1}(J^{k+1}_{N}) = \Delta^{k}(J^{k}_{N}) + \sigma^{k}_{0}\Delta\delta^{k}(J^{k}_{N})\ , \\ \nu^{k+1}(t^{0}) = -\sigma^{k}_{0}\ sign\ (a'_{sup}(t^{0})q^{k}_{1})\ . \end{cases}$$

Let situation (3.21) be represented .Then

$$Q^{k+1} = Q^{k+1}(J^{k+1}_{sup}\ ,\ L^{k+1}_{sup}\ ) =$$

$$= \begin{bmatrix} Q^{k}(J^{k}_{sup},L^{k}_{sup})+Q^{k}(J^{k}_{sup},L^{k}_{sup})P^{k}(L^{k}_{sup},j_{*})P^{k}(t^{0},J^{k}_{sup})Q^{k}(J^{k}_{sup},L^{k}_{sup}/w, & -Q^{k}(J^{k}_{sup},L^{k}_{sup})P^{k}(L^{k}_{sup},j_{*})/w \\ -P^{k}(t^{0},J^{k}_{sup})Q^{k}(J^{k}_{sup},L^{k}_{sup})/w\ , & 1/w \end{bmatrix} \qquad (3.24)$$

$$w = P^k(t^0,j_*) - P^k(t^0,J^k_{sup})Q^k(J^k_{sup},L^k_{sup})P^k(L^k_{sup},j_*) ;$$

$$P^k(t^0,J^k_{sup}) = (a_j(t^0),\ j\in J^k_{sup}) ;$$

$$P^k(t^0,j_*) = a_{j_*}(t^0) ;$$

$$\begin{cases} \nu^{k+1}(T^{k+1}_{sup}\setminus t^0) = \nu^k(T^k_{sup}) + \sigma^k_0\Delta\nu^k(T^k_{sup}) , \\ \mu^{k+1}(M) = \mu^k(M) + \sigma^k_0\Delta\mu^k(M) , \end{cases}$$

$$\begin{cases} \Delta^{k+1}(J^{k+1}_N) = \Delta^k(J^k_N\setminus j_*) + \sigma^k_0\Delta\delta^k(J^k_N\setminus j_*) , \\ \nu^{k+1}(t^0) = -\sigma^k_0\ sign\ (a'_{sup}(t^0)q^k_1) . \end{cases} \tag{3.25}$$

Proceed to *Step 6*.

Step 5. Introduce τ into the support. For this case

$$\Delta\lambda^{k'} = (\ \Delta\nu^k\ ,\ \Delta\mu^k\ )' = (\Delta\nu^k(T^k_{sup}),\ (\Delta\mu^k_j,\ j\in M))' =$$

$$= a'_{sup}(\tau)Q^k(J^k_{sup}\ ,\ L^k_{sup})\ sign\ (\xi^*(\tau) - \omega(\tau-h)) , \tag{3.26}$$

$$\Delta\delta^{k'} = \Delta\delta^{k'}(J) = \Delta\lambda^{k'}P^k(L^k_{sup},J) - a'(\tau)\ sign\ (\xi^*(\tau) - \omega(\tau-h))$$

Calculate according (3.10)-(3.12),(3.19) the value σ^k_0

Change the support $S^k_{sup} \longrightarrow S^{k+1}_{sup}$ according to (3.20),(3.21)

and evaluate Q^{k+1}, ν^{k+1}, μ^{k+1}, Δ^{k+1}_N following (3.22)-(3.27). Let $S^k_{sup} = S^{k+1}_{sup}$; $Q^k = Q^{k+1}$; $P^k = P^{k+1}$; $\nu^k = \nu^{k+1}$; $\mu^k = \mu^{k+1}$; $\Delta^k_N = \Delta^{k+1}_N$. We come back to *Step 2*.

Step 6. If $\rho_{k+1} > \xi^*(\tau)$ then the k-th iteration of the algorithm $C^k(\tau-h) \longrightarrow C^{k+1}(\tau-h)$ at the moment $\tau - h$ is completed.

If

$$\rho_{k+1} \le \xi^*(\tau) , \tag{3.27}$$

the functioning of the optimal estimator at the moment $\tau - h$ is finished ($k+1 = p$). The zero state of the algorithm at moment τ is

$$C^{0}(\tau) = (C^{k+1}(\tau-h)\backslash\rho_{k+1}) \cup (\rho_0 = \omega(\tau)).$$

The algorithm for the case $\omega(\tau-h) > \xi^*(\tau)$ is completely described. The case $\omega(\tau-h) < \xi_*(\tau)$ is analysed similarly.

Remark 3.1. While realizing *Step 5* in the recount formulae of potentials and estimates (3.23),(3.25) we set

$$sign(a'_{sup}(\tau)q_1^k) = \begin{cases} sign\ (\xi^*(\tau) - \omega(\tau-h)) & \text{, when } \omega(\tau-h) > \xi^*(\tau); \\ sign\ (\xi_*(\tau) - \omega(\tau-h)) & \text{, when } \omega(\tau-h) < \xi_*(\tau). \end{cases}$$

4.4. PROGRAM SOLUTION OF τ- CONTROL PROBLEM.

According to (1.12) the τ-a posteriori optimal control $\hat{u}^o(\cdot)$, $t=\tau,\tau+h,\dots,t^*-h$, is a solution of the extremal problem

$$h_o'x(t^*) \longrightarrow max, \quad x(t+h)=A(t+h)x(t)+b(t,h)u(t);$$

$$x(\tau)=0; \qquad h_i'x(t^*) \geq \hat{g}_i^\tau, \qquad i=\overline{1,m}; \tag{4.1}$$

$$u_*(t) \leq u(t) \leq u(t^*), \ \tau \leq t \leq t^*-h$$

where

$$\hat{g}_i^\tau = g_i - \hat{\gamma}_i^\tau - \sum_{t=0}^{\tau-h} h_i'x(t),$$

$\hat{\gamma}_i^\tau$ is the estimate of the i-th observation problem (2.7), $i=\overline{1,m}$.

The solution of terminal control problem (4.1) (see Chapter 1) is the totality of $\{ \hat{u}^o(\cdot|\tau), S_{sup}(\tau)\}$ where

$$S_{sup}(\tau) = \{ I_{sup}(\tau), T_{sup}(\tau) \}, \ I_{sup}(\tau) \subset I = \{1,2,\dots,m \},$$

$$T_{sup}(\tau) = \{ \tau_1,\dots, \tau_l\}, \ \tau \leq \tau_1(\tau) \leq \dots \leq \tau_l(\tau) \leq t^*-h.$$

Along with it relations

$$det\ P(\tau) = 0, \quad P(\tau) = \left[\begin{matrix} h_i'F_h(t^*,t)b(t,h), & t\in T_{sup}(\tau)) \\ i\in I_{sup}(\tau) & \end{matrix} \right]$$

hold true.

The vector of potentials

$$\upsilon'(\tau)=c'_{sup}Q(\tau),\ c_{sup}=(c(t),t\in T_{sup}(\tau)),$$

$$c(t)=h'_0F_h(t^*,t)b(t,h),\ \tau \le t \le t^*-h,\ Q(\tau) = P^{-1}(\tau)$$

corresponds to support $S_{sup}(\tau)$. The co-trajectory $\psi(t)=\psi(t|\tau)$, $\tau \le t \le t^*-h$, accompanying support $S_{sup}(\tau)$ as a solution of conjugate system

$$\psi'(t-h)=\psi'(t)A(t,h),\psi'(t^*-h)= h'_0- \upsilon'(I_{sup})H(I_{sup},J),$$

$$H(I_{sup},J)= \left[\begin{array}{c} h'_i(J) \\ i\in I_{sup}(\tau)\end{array}\right]$$

is constructed with the help of vector $\upsilon(\tau)$.

Co-trajectory generates the co-control

$$\Delta(t) = \Delta(t|\tau),\ \tau \le t \le t^*-h:\ \Delta(t) = -\psi'(t)b(t,h). \qquad (4.2)$$

We already know that the optimal control $\hat{u}^0(t),t \in T_N(\tau)= T(\tau)\backslash T_{sup}(\tau)$ at the non-supporting moments of time has the form

$$\hat{u}^0(t) \begin{cases} =u_*(t) \text{ when } \Delta(t)>0; \\ =u^*(t) \text{ when } \Delta(t)<0; \\ \in[u_*(t),u^*(t)] \text{ when } \Delta(t)=0,t\in T_N(\tau). \end{cases}$$

Without loss of generality we consider that for trajectory $x^0(t)$, $t\in T(\tau)$ generated by the control $\hat{u}^0(t)$, $t \in T(\tau)$, the equations

$$h'_ix(t^*) = \hat{g}^{\tau}_i,\ i\in I_{sup}\ (\upsilon(i)\le 0,i\in I_{sup})$$

are fulfilled.

The set of optimal control values $\hat{u}^0_{sup}= (\hat{u}^0(t),t\in T_{sup}(\tau))$ at the support moments is calculated according to

$$\hat{u}^0_{sup}=Q(\tau)g(\tau),\ g(\tau)= \left[\begin{array}{c} g_i(\tau) \\ i\in I_{sup}(\tau)\end{array}\right],$$

$$g_i(\tau) = \hat{g}_i^\tau - \sum_{t \in T_N(\tau)} H(I_{sup}, J) x_{\Delta^0}(t).$$

Later on we shall utilize the following information (see Chapter 3):

$$T_{N+}(\tau) = \{ t \in T_N(\tau) : \Delta(t)>0, \Delta(t)\Delta(t-h)<0 \} \cup$$

$$\cup \{ t \in T_N(\tau) : \Delta(t)>0, t - h \in T_{sup}(\tau)\},$$

$$T_{N-}(\tau) = \{ t \in T_N(\tau) : \Delta(t)<0, \Delta(t)\Delta(t-h) < 0 \} \cup$$

$$\cup \{ t \in T_N(\tau) : \Delta(t)<0, t - h \in T_{sup}(\tau) \},$$

$$F(t^*, t), \ t \in T_{sup}(\tau) \cup \tau_N \cup (t^*-h).$$

4.5. OPTIMAL CONTROLLER SYNTHESIS.

We describe briefly the optimal controller algorithm.

We call the array

$$C^k(\tau) = \{ u^{(k)}(t);\ t\in T(\tau+h),\ w^k;\ S^k_{sup} = \{ I^k_{sup}, T^k_{sup} \} ;\ T^k_{N+} ;$$

$$T^k_{N-} ;\ \Delta g^k ;\ \Phi^k(t),\ t\in T^k_{sup} \cup \tau_N \cup (t^*-h) ;\ \psi^k(t),\ t\in T^k_{N+} \cup$$

$$\cup\ T^k_{N-} \cup (t^*-h) ;\ \nu^k ;\ Q^k \}$$

the state of the algorithm on the k-th iteration at the moment τ. As the initial state $C^0(\tau)$ at the moment τ we choose an array with components

$$u^{(0)}(t) = \hat{u}^0(t|\tau),\ t\in T(\tau+h) ;\ w^0 = Hx^0(t^*) - \hat{g}^\tau ;$$

$$S^0_{sup} = S_{sup}(\tau) = \{ I_{sup}(\tau),\ T_{sup}(\tau) \} ;\ T^0_{N+} = T_{N+}(t) ;$$

$$T^0_{N-} = T_{N-}(\tau) ;\Delta g^0 = \hat{\gamma}^{\tau-2h} - \hat{\gamma}^{\tau-h} ;\Phi^0(t) = F(t^*,t),t\in T_{sup}(\tau) \cup$$

$$\cup\ \tau_N \cup (t^*-h) ;\ \nu^0 = \nu(\tau) ;\ Q^0 = Q(\tau).$$

Iteration of the algorithm consists of the following steps.

Step 1. If $l = 0$, proceed to *Step 2*. Let $l \neq 0$. Compare τ with τ_1. If $\tau < \tau_1$, proceed to *Step 8*.

Step 2. Calculate vectors

$$\Delta u^k(T^k_{sup}) = Q^k\Delta g^k(I^k_{sup}) ;\ \Delta u^k(T^k_N) = 0 ,\ T^K_N = T(\tau+h)\backslash T^k_{sup};$$

$$\Delta w^k(I^k_N) = \sum_{t\in T^k_{sup}} H(I^k_N,\ J)\Phi^k(t)b(t,\ h)\Delta u^k(T) ,$$

$$\Delta w^k(I^k_{sup}) = 0 ,\ I^k_N = I\backslash I^k_{sup} .$$

Step 3. Calculate numbers a^k , β^k , θ^k :

$$\alpha^k = \alpha(\tau_s) = \min \alpha(t) ,\ t\in T^k_{sup} :$$

$$\alpha(t) = \begin{cases} (u_*(t) - u^{(k)}(t))/\Delta u^k(t) \text{ , when } \Delta u^{(k)}(t) < 0 \text{ ;} \\ (u^*(t) - u^{(k)}(t))/\Delta u^{(k)}(t) \text{ , when } \Delta u^{(k)}(t) > 0 \text{ ;} \end{cases}$$

$$\beta^k = \beta(i_0) = min\ \beta(i),\ i\in I^k_N \ :$$

$$\beta(i) = \begin{cases} -w^k_i/(\Delta w^k_i - \Delta g^K_i) \text{ , when } (\Delta w^k_i - \Delta g^k_i) < 0 \text{ ;} \\ \infty \text{ , when } (\Delta w^k_i - \Delta g^k_i) \geq 0 \text{ .} \end{cases}$$

Let $\theta^k = min\{ 1,\ \alpha^k,\ \beta^k \}$. If $\theta^k = 1$, proceed to *Step 4*. At $\theta^k < 1$, proceed to *Step 5*.

Step 4. Let $\hat{u}^0(\tau+h|\tau+h) = u^k(\tau+h) + \Delta u^k(\tau+h)$. If $\tau+h = t^*-lh$,the algorithm stops:

$$\hat{u}^0(\tau+ih|\tau+ih) = u^{(k)}(\tau+ih)+\Delta u^{(k)}(\tau+ih),\ i =\overline{i,l}.$$

At $\tau+h < t^*-lh$, we construct the initial state $C^0(\tau+h)$ for the moment $\tau + h$:

$$u^{(0)}(t) = u^{(k)}(t) + \Delta u^k(t) \text{ , } t\in T(\tau+2h) \text{ ; } w^0 = w^k + \Delta w^k \text{ ;}$$

$$S^0_{sup} = S^k_{sup} \text{ ; } T^0_{N+} = T^k_{N+} \text{ ; } T^0_{N-} = T^k_{N-} \text{ ; } \Delta g^0 = \hat{\gamma}^{\tau-h} - \hat{\gamma}^{\tau} \text{ ;}$$

$$\Phi^0(t) = \Phi^k(t) \text{ , } t\in T^0_{sup} \cup \tau_N \cup (t^*-h) \text{ ; } \psi^0(t) = \psi^k(t) \text{ ,}$$

$$t\in T^0_{N+} \cup T^0_{N-} \cup (t^*-h) \text{ ; } \nu^0 = \nu^k \text{ ; } Q^0 = Q^k.$$

Step 5. Calculate

$$\Delta^k(t) = -\psi^{k\prime}(t)b(t,\ h),\ t\in T^k_{N+} \cup T^k_{N-} \cup (t^*-h).$$

In case $\theta^k = \beta^k = \beta(i_0)$ we set

$$\mu^k(i) = [\ h'_{i_0}\ \Phi^k(t)b(t,\ h),\ t\in T^k_{sup}\ ]Q^k(T^k_{sup} \text{ , } i) \text{ , } i\in I^k_{sup} \text{ ;}$$

$$\xi^k(t) = [\, h'_{i_0} - \mu^{k\prime}(I^k_{sup})H(I^k_{sup},J)]\Phi^k(t),\ t\in T^k_{N+} \cup T^k_{N-} \cup (t^*-h),$$

$$\Phi^k(t) = A^{-1}(\theta,h)\Phi^k(\theta)\ ,\ t = \theta+h\ ,\ \theta\in T^k_{sup}\ .$$

In the case $\theta^k = \alpha^k = \alpha(\tau_s)$ we have

$$\mu^k(I^k_{sup}) = \rho q(\tau_s),\ \xi^k(t) = \rho q(\tau_s)H(I^k_{sup},J)\Phi^k(t),\ t\in T_{N+} \cup T^k_{N-}\cup$$

$$\cup\ (t^*-h)\ ,\quad \Phi^k(t) = A^{-1}(\theta,\ h)\Phi^k(\theta)\ ,\ t = \theta+h,\ \theta\in T^k_{sup}\ ,$$

$$\rho = -sign\ \Delta u^k(\tau_s)\ .$$

Calculate $\delta^k(t) = \xi^{k\prime}(t)b(t,\ h)\ ,\ t\in T_{N+} \cup T^k_{N-} \cup (t^*-h).$
Proceed to *Step 6*.

Step 6. Calculate

$$\sigma^k = min\ [\ \sigma(t),\ t\in T_{N+} \cup T^k_{N-} \cup (t^*-h)\ ;$$

$$\omega(i),\ i\in I^k_{sup}\ ]\ ;\ s(t),\ t\in T_{N+} \cup T^k_{N-}\ :$$

$\sigma(t) = -\Delta^k(t)/\delta^k(t),\ s(t) = 0$ either $t\in T^k_{N+}\ ,\ \delta^k(t) < 0$ or

$t\in T^k_{N-}\ ,\ \delta^k(t) > 0\ ;\ \sigma(t) = -\Delta^k(t-h)/\delta^k(t-h) =$

$= -\ \psi^{k\prime}(t)A(t,\ h)b(t-h,\ h)/\xi^{k\prime}(t)A(t,\ h)b(t-h,\ h),\ s(t) = h,$

either $t\in T^k_{N+}\ ,\ \delta^k(t-h) > 0$ or $t\in T^k_{N+}\ ,\ \delta^k(t-h) > 0$ or

$t\in T^k_{N-},\ \delta^k(t-h) < 0,\ (t-h)\notin T^k_{sup};\ \sigma(t) = -\Delta^k(t-2h)/\delta^k(t-2h) =$

$= -\ \psi^{k\prime}(t)A(t,h)A(t-h,h)b(t-2h,h)/\xi^{k\prime}(t)A(t,h)A(t-h,h)b(t-2h,h),$

$s(t) = 2h,$ either $t\in T^k_{N+},\ \delta^k(t-2h) > 0$

or $t\in T^k_{N-}\ ,\ \delta^k(t-2h) < 0\ ,\ (t-2h)\notin T^k_{sup}\ ;\ (t-h)\in T^k_{sup}\ ;$

$\sigma(t^*-h) = -\Delta^k(t^*-h)/\delta^k(t^*-h)$, when

$\Delta^k(t^*-h)\delta^k(t^*-h) < 0\ ,\ (t^*-h)\notin T^k_{sup}\ ,$

$$\sigma(t) = \infty \text{ in other cases;}$$

$$\omega(i) = -\nu^k(i)/\mu^k(i), \text{ either } \nu^k(i)\mu^k(i) < 0$$

or $\nu^k(i) = 0$, $\mu^k(i) > 0$; $\omega(i) = \infty$ in other cases.

Proceed to *Step 7*.

Step 7. Transform the set $S^k_{sup} = \{I^k_{sup}, T^k_{sup}\}$ and the matrix Q^k by a standard way with the help of formulae from Sections 3.1 and 3.3.

Let

$$u^{(k+1)}(t) = u^{(k)}(t) + \theta^k \Delta u^k(t) \ , \ t \in T(\tau+h) \ ; \ \Delta g^{k+1} =$$

$$= (1-\theta^k)\Delta g^k \ ; \quad \psi^{k+1}(t) = \psi^k(t) + \sigma^k \xi^k(t) \ , \ t \in T^k_{N+} \cup T^k_{N-} \cup$$

$$\cup \ (t^* - h) \ ; \ \nu^{k+1} = \nu^k + \sigma^k \mu^k \ ; \ w^{k+1} = w^k + \theta^k \Delta w^k \ .$$

Proceed to *Step 2*.

Step 8. Let $s = 1$, $\theta^0 = \alpha^0 = \alpha(\tau_1) = 0$, $\Delta u^0(\tau_1) =$ $= g(\tau_1)\Delta g^0(I^0_{sup})$. We come back to *Step 5*.

Example 5.1. Illustrate the results by an example.

It is necessary to transfer a material point which begins to move along a rectilinear path from some neighbourhood of the given point to a certain region and provide at the moment a velocity the guaranteed value of which is maximal.

It is also necessary to take into account that all information about the control comes from the device making measurements of the summarized value of a position and a velocity with limited exactness.

The mathematical model of problem has the form

$$x_2(3) \longrightarrow \max \ , \quad x_1(t+h) = hx_2(t) \ ,$$

$$x_2(t+h) = x_2(t) + hu(t) \ , \ |x_1(0)| \le 1 \ , \ x_2(0) = 0 \ ,$$

$$x_1(3) \le 1 \ , \quad 0 \le u(t) \le 1 \ , \ t = 0. \ h, \ 2h, \ \ldots \ , \ 3 \ ;$$

$$h = 0.5 \ ; \quad y = x_1 + x_2 + \xi \ , \quad 0 \le \xi \le 1 \ .$$

We shall present the results of functioning the optimal estimator and the controller for the case where the point began to move from the point $x_1(0) = 0$ and the following measurement errors were realized:

$\xi(0) = 1/2$, $\xi(0.5) = 1/4$, $\xi(1) = 1/2$, $\xi(1.5) = 1/5$, $\xi(2) = 1/4$,

but this information was not known either by the estimator or the controller.

The a priori optimal control $\check{u}^o(\cdot)$ constructed at the moment $t = 0$ without results of observation has the form represented in Fig. 5.1.

The guaranteed value of the quality criterion is equal to $J(\check{u}^o(\cdot)) = 1/2$.

If the initial state $x_1(0) = 0$ is known for the controller at the moment $t = 0$ then the optimal control $u^o(\cdot)$ has the form presented in Fig.5.2. The value of the quality criterion would reach the number $J(\check{u}^o(\cdot)) = 9/4$.

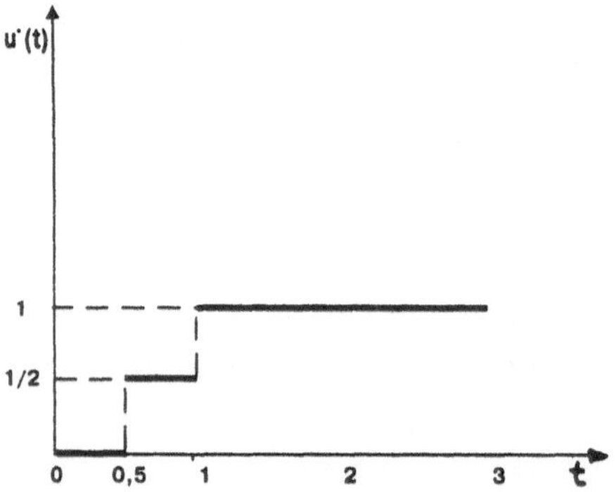

Fig.5.1

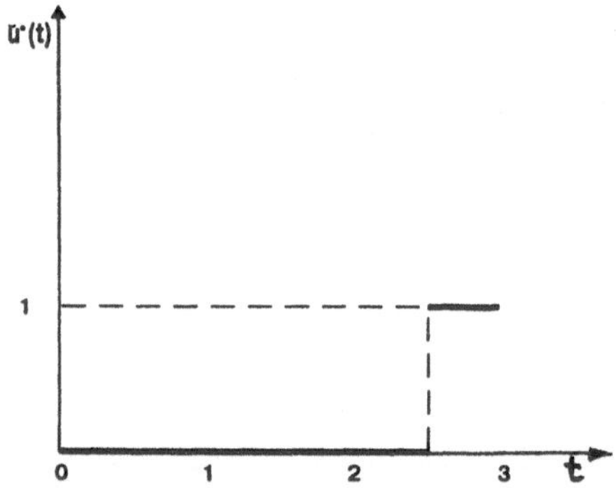

Fig.5.2

After processing the signal $y(0)=1/2$ by the estimator the controller produced the control $\hat{u}^o_1(\cdot)$, presented inFig.5.3. $(J(\hat{u}^o_1(\cdot)) = 3/4)$.

Performing by analogy the processing of signals $y(0.5) = 1/4$, $y(1) = 1/2$, $y(1.5) = 1/5$, $y(2) = 1/4$, the controller constructed a priori optimal controls presented in Fig.5.4 $(\hat{u}^o(\cdot) = \hat{u}^o_2(\cdot) = \hat{u}^o_3(\cdot) = \hat{u}^o_4(\cdot))$. It is clear that the processing of measurements $y(t)$, $t \geq 1$, does not influence control.

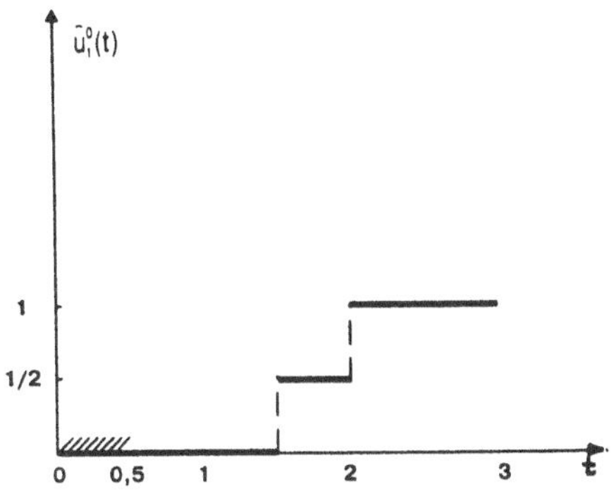

Fig.5.3

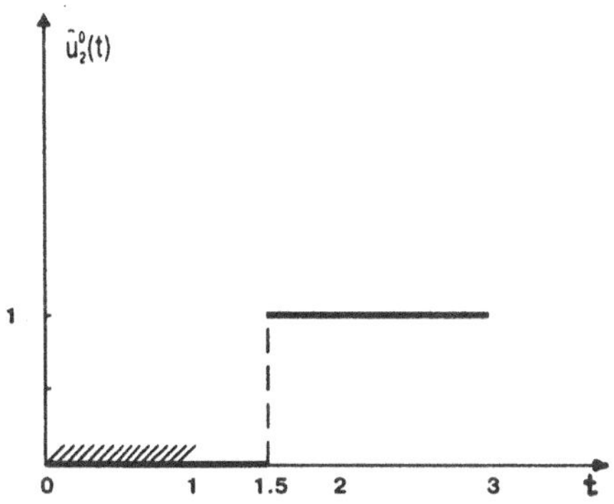

Fig.5.4

The value of the quality criterion for the constructed control is equal to $J(\hat{u}^o(\cdot)) = 3/2$.

The value $J(\hat{u}^o(\cdot)) - J(\check{u}^o(\cdot)) = 1$ characterizes the increase in control efficiency at the expense of the measuring device.

The loss of efficiency due to the errors of the measuring device equals $J(u^o(\cdot)) - J(\hat{u}^o(\cdot)) = 3/4$.

APPENDIX

ADAPTIVE METHOD OF LINEAR PROGRAMMING

1. LINEAR PROGRAMMING PROBLEM.

We shall be concerned with the classical problem of linear programming (LP)

maximize

$$F(x_1,x_2,\dots,x_n) = c_1x_1+\dots+c_nx_n$$

subject to

$$\begin{aligned} a_{11}x_1 + a_{12}x_2 +\dots+ a_{1n}x_n &= b_1, \\ a_{21}x_1 + a_{22}x_2+\dots+ a_{2n}x_n &= b_2, \\ a_{n1}x_1 + a_{n2}x_2+\dots+a_{nn}x_n &= b_n, \end{aligned} \qquad (1.1)$$

$$d_{*1} \le x_1 \le d_1^*,\ d_{*2} \le x_2 \le d_2^*,\ \dots,\ d_{*n} \le x_n \le d_n^*.$$

Let $J = \{1,2,\dots, n\}$ be an index set of the variables $x_1,x_2,\dots,x_n$ and $I = \{1,2,\dots,m\}$ be an index set of the parameters $b_1, b_2, \dots, b_m$. Introduce the vectors $x = x(J) = (x_j, j \in J) = (x_1, x_2, \dots, x_n)$, $c = c(J)$, $b = b(I)$, $d_* = d_*(J)$, $d^* = d^*(J)$ and the matrix $A = A(I,J) = (a_{ij}, i \in I, j \in J)$. Then problem (1.1) takes the form

$$F(x) = c'x \longrightarrow max,$$
$$Ax = b, \; d_* \le x \le d^*. \qquad (1.2)$$

Denote the feasible region as

$$X = \{ \; x \in R^n \colon \; Ax = b, \; d_* \le x \le d^* \; \}.$$

Below in Sections 1-4 we shall assume

$$rank \;\; A = m.$$

Elements of the set X are said to be feasible points. The feasible point satisfies both the general ($Ax = b$) and the simple ($d_* \le x \le d^*$) constraints.

Definition 1.1. The feasible point x° at which the objective function achieves maximal value is named a solution to the problem or an optimal feasible point.

Definition 1.2. For a given value $\varepsilon \ge 0$ the feasible point x^{ε} is called ε-optimal (suboptimal) if it satisfies the inequality

$$F(x^{\circ}) - F(x^{\varepsilon}) \le \varepsilon.$$

Definition 1.3. The set of m indices $J_{sup} \subset J$ is called a *support* if the matrix $P = A(I, J_{sup})$ is non-singular.

The support J_{sup} is a minimal set of indexes of J such that for any choice of the vector b and components $x_N = (\; x_j, \; j \in J_N \;), \; J_N = J \backslash J_{sup}$, the general constraints can be satisfied by choosing the components $x_{sup} = (\; x_j, \; j \in J_{sup})$.

Really, the general constraints $Ax = b$ in component form can be expressed as

$$Px_{sup} + A(I, J_N)x_N = b.$$

Therefore, to satisfy the equality it is sufficient to set

$$x_{sup} = P^{-1}(b - A_N x_N), \; A_N = A(I, J_N). \qquad (1.3)$$

Because of a special role played by the inverse matrix with regard to the support, we introduce the notation

$$Q = Q(J_{sup}, I) = P^{-1}.$$

In the iteration method the support will be changed together with feasible points. Therefore the main object of transformation is a pair comprising a feasible point and a support.

Definition 1.4. The pair $\{ x, J_{sup} \}$ comprising a feasible point and a support will be called a *support feasible* (SF-) *point*. The SF-point will be called non-degenerate if

$$d_{*j} < x_j < d_j^*, \; j \in J_{sup}.$$

2. OPTIMALITY CRITERION.

We shall assume that the initial SF-point is known. Wefirst examine the behaviour of the objective function when the feasible point is changed.

2.1. Formula of the objective value increment.

On a level with the SF-point $\{x, J_{sup}\}$ we consider an arbitrary n-vector $\bar{x}$ satisfying the constraints $A\bar{x} = b$. Such vectors are called pseudo-feasible points. We set $\Delta x = \bar{x} - x$ and calculate the increment of the objective value

$$\Delta F(x) = F(\bar{x}) - F(x) = c'\bar{x} - c'x = c'\Delta x. \tag{2.1}$$

It is obvious that

$$A\Delta x = A(\bar{x} - x) = A\bar{x} - Ax = b - b = 0 \tag{2.2}$$

or

$$P\Delta x_{sup} + A_N \Delta x_N = 0$$

and

$$\Delta x_{sup} = -QA_N \Delta x_N. \tag{2.3}$$

Thus for any $\Delta x_N = (\Delta x_j,\ j \in J_N)$ and Δx_{sup} defined by (2.3) we obtain the vector $\Delta x = (\Delta x_{sup}, \Delta x_N)$ such that $\bar{x} = x + \Delta x$ satisfies the general constraints.

Substituting the vector Δx into (2.1) we get

$$\Delta F(x) = c'_{sup}\Delta x_{sup} + c'_N x_N = -(c'_{sup} QA_N - c')\Delta x_N. \tag{2.4}$$

The vector

$$\Delta' = (\Delta_j, j \in J) = c'_{sup} QA - c' \tag{2.5}$$

will be called a support gradient. It is evident that its support components are equal to zero:

$$\Delta'_{sup} = c'_{sup}QP - c'_{sup} = c'_{sup} - c'_{sup} = 0.$$

To calculate non-support components of a support gradient it is convenient to use the vector of multipliers

$$u' = c'_{sup}Q. \tag{2.6}$$

Then we have

$$\Delta'_N = u'A_N - c'_N \tag{2.7}$$

or

$$\Delta_j = u'a_j - c_j, \quad j \in J_N \tag{2.8}$$

where $a_j = A(I,j)$ is the j-th column of the matrix A. From (2.4) and (2.5) we obtain

$$\Delta F(x) = -\Delta'_N \Delta x_N = -\sum_{j \in J_N} \Delta_j \Delta x_j. \tag{2.9}$$

Using (2.9) we shall clarify a physical sense of the support gradient Δ. Let $\Delta x_N = (0, \ldots, 0, \Delta x_k, 0, \ldots, 0)$, where $k \in J_N$ is a certain index. The components Δx_{sup} are found from (2.3):

$$\Delta x_{sup} = -QA_N \Delta x_N = -Qa_k \Delta x_k.$$

According to (2.9)

$$\Delta F(x) = -\Delta_k \Delta x_k.$$

Remembering definition of derivative and its mechanical sense, we come to the conclusion that Δ_k is the rate of change of the objective function taken with opposite sign when the k-th non-support component of the feasible point x is increased and all the non-support components (besides the k-th) are fixed. At the same time the support components are changed in such a way to satisfy the general constraints.

The k-th component of the "classical" derivative of the objective function is equal to $\partial F/\partial x_k = c_k$. In contrast with c_k, $-\Delta_k$ is a conditional derivative, calculated under conditions of fulfilment of the general constraints. In our case the constraints are satisfied with the help of the support. Therefore $-\Delta_k$ can be called a support derivative.

2.2. The optimality criterion.

Let $\{x, J_{\text{sup}}\}$ be an SF-point. Calculate the support gradient (2.8).

Theorem 2.1. For a feasible point x to be optimal it is sufficient and, in the case of non-degeneracy of the SF-point $\{x, J_{\text{sup}}\}$, also necessary, that

$$\Delta_j \geq 0 \text{ for } x_j = d_{*j}, \ \Delta_j \leq 0 \text{ for } x_j = d_j^*,$$
$$\Delta_j = 0 \text{ for } d_{*j} < x_j < d_j^*, \ j \in J_N. \tag{2.10}$$

The proof of Theorem 2.1 can be obtained using (2.9).

Definition 2.1. The pair $\{x^0, J^0_{\text{sup}}\}$ satisfying the relations (2.10) will be called an optimal SF-point.

We note here that the optimality of the feasible point can not be identified if it is examined with an unfit support.

3. SUBOPTIMALITY CRITERION.

In applied problems it is often sufficient to have suboptimal feasible points. For this reason the problem of identification of ε-optimal feasible points appears. The suboptimality criterion allows us to identify an ε-optimal feasible point and to solve the problem without additional expenditures which are required to obtain the result with strict accuracy.

3.1. The suboptimality estimate. Sufficient condition of suboptimality.

In the set of pseudo-feasible points we consider the subset consisting of the vectors $\bar{x}$, non-support components of which satisfy the simple constraints

$$d_{*j} \le \bar{x}_j = x_j + \Delta x_j \le d_j^*,\ j \in J_N .$$

The maximal value (3.1) in the subset is equal to

$$\max \Delta F(x) = \max_{\substack{d_{*j}-x_j \le \Delta x_j \le d_j^*-x_j,\\ j\in J_N}} \left(-\sum_{j\in J_N} \Delta_j \Delta x_j\right) =$$

$$= \sum_{j\in J_N} \left(\max_{d_{*j}-x_j \le \Delta x_j \le d_j^*-x_j} (-\Delta_j \Delta x_j)\right) =$$

$$= \sum_{\substack{\Delta_j>0,\\ j\in J_N}} \Delta_j(x_j-d_{*j}) + \sum_{\substack{\Delta_j<0,\\ j\in J_N}} \Delta_j(x_j-d_j^*)$$

The value

$$\beta = \beta(x, J_{sup}) = \sum_{\substack{\Delta_j > 0, \\ j \in J_N}} \Delta_j (x_j - d_{*j}) + \sum_{\substack{\Delta_j < 0, \\ j \in J_N}} \Delta_j (x_j - d_j^*) \qquad (3.1)$$

will be called a suboptimality estimate of the SF-point $\{x, J_{sup}\}$ since

$$F(x^0) - F(x) \le \beta(x, J_{sup}). \qquad (3.2)$$

The suboptimality estimate is finite if $d_{*j} > -\infty$ for $\Delta_j > 0$ and $d_j^* < +\infty$ for $\Delta_j < 0$. We shall consider only problems when $\beta(x, J_{sup}) < \infty$.

Theorem 3.1. (Sufficient condition of suboptimality). For a feasible point x to be ε-optimal for given $\varepsilon \ge 0$ it is sufficient for there to exist a support J_{sup} such that

$$\beta(x, J_{sup}) \le \varepsilon.$$

The proof follows from (3.2).

3.2. The dual problem. Accompanying co-point. Coordinated dual point. Accompanying pseudo-feasible point.

Let $\{x^0, J^0_{sup}\}$ be an SF-point satisfying the optimality criterion (2.10) (such a point can be constructed by the algorithm described below). Let u^0 be the vector of multipliers $u^{0\prime} = c'_{sup} Q$ corresponding to the support J^0_{sup}.

Using (2.8),(2.10) we get

$$a'_j u^0 - c_j \ge 0 \quad \text{for } x_j = d_{*j},$$
$$a'_j u^0 - c_j \le 0 \text{ for } x_j = d_j^*, \qquad (3.3)$$
$$a'_j u^0 - c_j = 0 \quad \text{for } d_{*j} < x_j < d_j^*, \ j \in J_N.$$

From (3.4) we have

$$a'_j u^0 - c_j = 0, \ j \in J^0_{sup}. \qquad (3.4)$$

We introduce the vector

$$\delta^o = \delta^o(J) = A'u^o - c. \tag{3.5}$$

It follows from (3.4) that

$$\delta^o_{sup} = 0, \ \delta^{o\prime}_{N} = c'_{sup}QA_N - c'_N. \tag{3.6}$$

Construct the vectors v^o, w^o with

$$v^o_j = \delta^o_j, \ w^o_j = 0 \quad \text{for} \quad \delta^o_j \geq 0;$$
$$v^o_j = 0, \ w^o_j = -\delta^o_j \quad \text{for} \quad \delta^o_j < 0, \ j \in J. \tag{3.7}$$

According to (3.5)-(3.7) the collection $\lambda^o = (\ y = u^o,\ v = v^o,\ w = w^o)$ satisfies the relations

$$A'y - v + w = c, \quad v \geq 0, \ w \geq 0. \tag{3.8}$$

We calculate the value

$$\Phi(\lambda^o) = b'u^o - d'_*v^o + d^{*\prime}w^o = c'_{sup}Qb -$$

$$- \sum_{\delta^o_j \geq 0,\, j\in J_N} d_{*j}\delta^o_j - \sum_{\delta^o_j < 0,\, j\in J_N} d^*_j\delta^o_j =$$

$$= c'_{sup}Qb - \sum_{j\in J_N} x^o_j\delta^o_j = c'_{sup}Qb - (c'_{sup}QA_N - c'_N)x^o_N =$$

$$= c'_{sup}(Qb - QA_N x^o_N) + c'_N x^o_N = c'_{sup}x^o_{sup} + c'_N x^o_N = c'x^o.$$

Thus

$$\Phi(\lambda^o) = F(x^o). \tag{3.9}$$

Let $\lambda = (\ y,\ v,\ w\ )$ be a collection satisfying (3.10) and x be an arbitrary feasible point. Estimate the value $\Phi(\lambda)$:

$$\Phi(\lambda) = b'y - d_*'v + d^{*\prime}w \geq x'A'y - x'v + x'w =$$

$$x'(A'y - v + w) = c'x$$

i.e.

$$\Phi(\lambda) \geq F(x) \tag{3.10}$$

for any feasible point x and, in particular, for an optimal point:

$$\Phi(\lambda) \geq F(x^o). \tag{3.11}$$

Comparing (3.9) and (3.8) we conclude that $\lambda^o = (y^o, v^o, w^o)$ is a solution of the following extremal problem:

$$\begin{gathered} \Phi(\lambda) = b'y - d_*'v + d^{*\prime}w \longrightarrow min, \\ A'y - v + w = c, \ v \geq 0, \ w \geq 0. \end{gathered} \tag{3.12}$$

It is an LP problem. Just as problem (1.2) it is formed from the same parameters $\{c, A, b, d_*, d^*\}$. In the following, problem (1.2) will be called primal, and problem (3.12) will be called dual.

Definition 3.2. A collection $\lambda = (y,v,w)$ satisfying all the constraints of the dual problem (3.14) is called a dual feasible point. The solution $\lambda^o = (y^o, v^o, w^o)$ of problem (3.12) is an optimal dual feasible point.

The special solution $\lambda^o = (y^o, v^o, w^o)$ of the dual problem was constructed above for the support J^o_{sup}. We introduce a similar construction for an arbitrary support J_{sup}. Form $u' = c'_{sup}Q$. The vector $\delta = A'u - c$ will be called a co-point accompanying the support J_{sup} (or an accompanying co-point). The accompanying co-point coincides with the support gradient, i.e. $\delta = \Delta$. Using the co-point δ we construct v, w with

$$v_j = \delta_j, \quad w_j = 0 \quad \text{for} \quad \delta_j \geq 0;$$

$$v_j = 0, \quad w_j = -\delta_j \quad \text{for} \quad \delta_j < 0, \ j \in J. \qquad (3.14)$$

The collection $\lambda = (y,v,w)$ satisfies the constraints of the dual problem (3.14). It will be called an accompanying dual feasible point.

Let y be an arbitrary m-vector. Construct the co-point $\delta = A'y - c$ and the vectors v,w according to (3.14). The vector $\lambda = (y, v, w)$ is a dual feasible point. The dual feasible point satisfying (3.16) is called a coordinated dual feasible point. By construction any accompanying dual feasible point is coordinated and $\delta_{sup} = 0$ for it.

The coordinated dual feasible points have an important extremal property. Let $\lambda = (y, v, w)$ be a coordinated dual feasible point and $\bar{\lambda} = (y, \bar{v}, \bar{w})$ be an uncoordinated one. Then $\Phi(\lambda) \leq \Phi(\bar{\lambda})$.

In the following, we shall consider coordinated dual feasible points only.

Let J_{sup} be a support and δ be an accompanying co-point. The vector $\kappa = \kappa(J) = (\kappa_{sup}, \kappa_N)$ with

$$\kappa_j = d_{*j} \quad \text{if} \quad \delta_j > 0, \quad \kappa_j = d_j^* \quad \text{if} \quad \delta_j < 0,$$

$$\kappa_j = d_{*j} \quad \text{or} \quad d_j^* \quad \text{if} \quad \delta_j = 0, \ j \in J_N; \qquad (3.15)$$

$$\kappa_{sup} = Q(b - A_N \kappa_N)$$

will be called an accompanying pseudo-feasible point.

Setting $\kappa_j = d_{*j}$ for $\delta_j = 0$ in (3.15) we believe that δ_j is infinitesimal positive number, i.e. $\delta_j = +0$; and in the case $\kappa_j = d_j^*$ for $\delta_j = 0$ we believe that δ_j is infinitesimal negative, i.e. $\delta_j = -0$.

The index set for which $\kappa_j = d_{*j}$ we denote by J_N^+ and $J_N^- = \{ j \in J_N : \kappa_j = d_j^* \}$.

An accompanying pseudo-feasible point is an optimal solution to the problem

$$c'x \longrightarrow \max,$$
$$Ax = b, \quad d_{*N} \le x_N \le d^*_N. \tag{3.16}$$

The conditions (3.15) fulfil the optimality conditions (2.10). If

$$d_{*sup} \le x_{sup} \le d^*_{sup}, \tag{3.17}$$

i.e. κ is a feasible point of the primal problem (1.2), then it is an optimal solution.

Thus the vector $\Delta x_N = \kappa_N - x_N$ maximizes the increment of the objective value when the suboptimality estimate is calculated and

$$\beta(x,J_{sup}) = \Delta'_N(x_N - \kappa_N). \tag{3.18}$$

For any support we have

$$c'\kappa = \Phi(\lambda) \tag{3.19}$$

where λ is the dual feasible point accompanying the support J_{sup} and κ is an accompanying pseudo-feasible point.

3.3. Decomposition of the suboptimality estimate. Degrees of non-optimality of a feasible point and a support.

Let $\{x,J_{sup}\}$ be an SF-point, and $\beta(x,J_{sup})$ be its suboptimality estimate calculated according to (3.1), $\lambda = (y, v, w)$ be the accompanying dual feasible point. Then

$$\beta(x,J_{sup}) = \Delta'_N(x_N - \kappa_N) = \delta'(x - \kappa) = (u'A - c')(x - \kappa) =$$

$$= c'\kappa - c'x = \Phi(\lambda) - F(x) = \Phi(\lambda) - \Phi(\lambda^o) + F(x^o) - F(x) =$$

$$= \beta(J_{sup}) + \beta(x),$$

i.e.

$$\beta(x, J_{sup}) = \beta(x) + \beta(J_{sup}) \tag{3.20}$$

where $\beta(x) = F(x^o) - F(x)$ is a deviation of the value $F(x)$ from the optimal and $\beta(J_{sup}) = \Phi(\lambda) - \Phi(\lambda^o)$ is a deviation of the value $\Phi(\lambda)$ from the optimal. The value $\beta(x)$ is called the degree of non-optimality of the feasible point, and $\beta(J_{sup})$ is the degree of non-optimality of the support.

It follows from (3.20) that $\beta(x, J_{sup})$ is an accurate suboptimality estimate of the feasible point x if

$$\beta(J_{sup}) = 0.$$

Just as the feasible point x^o is called optimal for $\beta(x^o) = 0$, the support J^o_{sup} will be called optimal if $\beta(J^o_{sup}) = 0$. Every support which together with some feasible point satisfies (2.10) is optimal.

3.4. The necessary condition of suboptimality.

From (3.20) we have the following.

Theorem 3.2. (The necessary condition of suboptimality). For the feasible point x to be ε-optimal for given $\varepsilon \geq 0$ it is necessary for there to exist a support J_{sup} such that inequality

$$\beta(x, J_{sup}) \leq \varepsilon$$

holds.

4. ITERATION OF THE ALGORITHM.

During the algorithm iterations the transfer $\{x, J_{sup}\} \longrightarrow \{\bar{x}, \bar{J}_{sup}\}$ from one SF-point to a new one is carried out. At the beginning we assume that an initial SF-point and $\varepsilon \geq 0$ are known. The construction of the initial SF-point (Phase I) will be considered in Section 5.

Iteration of the adaptive method is based on the principle of decreasing the suboptimality estimate. In other words the transfer $\{x, J_{sup}\} \longrightarrow \{\bar{x}, \bar{J}_{sup}\}$ is carried out in such a way that

$$\beta(\bar{x}, \bar{J}_{sup}) \leq \beta(x, J_{sup}).$$

The principle will be realized as two procedures making up the iteration: i) transformation of the feasible point $x \longrightarrow \bar{x}$ which decreases the degree of non-optimality of the feasible point: $\beta(\bar{x}) \leq \beta(x)$; ii) support variation of the $J_{sup} \longrightarrow \bar{J}_{sup}$ which decreases the degree of non-optimality of the support: $\beta(\bar{J}_{sup}) \leq \beta(J_{sup})$.

4.1. The transformation of the feasible point.

Let $\{x, J_{sup}\}$ be an SF-point. Calculate the support gradient $\Delta = A'u - c$ ($u = Q'c_{sup}$), the non-support components of accompanying pseudo-feasible point κ_N

$$\kappa_j = \begin{cases} d_{*j} & \text{for } j \in J_N^+ \\ d_j^* & \text{for } j \in J_N^- \end{cases} \tag{4.1}$$

and the suboptimality estimate

$$\beta(x, J_{sup}) = \Delta_N'(x_N - \kappa_N).$$

If $\beta(x,J_{sup}) \le \varepsilon$, then x is an ε-optimal feasible point and the solution process is stopped. Let $\beta(x,J_{sup}) > \varepsilon$. We construct

$$\kappa_{sup} = Q(b - A_N x_N)$$

and find the value μ of the maximal violation of simple constraints at vector κ_{sup}:

$$\mu = \max_{j \in J_{sup}} \rho(\kappa_j, [d_{*j}, d_j^*]).$$

If $\mu = 0$ ($d_{*j} \le \kappa_{sup} \le d_j^*$) then κ is an optimal feasible point. In this case solution process is also stopped.

If $\mu \le \mu_o$, where μ_o is the parameter of the method,we pass to changing the support (see below). Otherwise we transform the feasible point x.

We construct the direction

$$l = \kappa - x \tag{4.2}$$

and look for a new feasible point $\bar{x}$ among vectors

$$x(\theta) = x + \theta l, \ \theta \ge 0$$

where θ is a step length along l.

The main specific feature of the iteration is the method of constructing l. According to (4.2) the vector l is directed from x to κ which is an accompanying pseudo-feasible point, and satisfies the optimality conditions (2.10) with respect to non-support components, i.e. it is a solution of problem (3.18). It has property $\beta(\kappa,J_{sup}) = 0$ and the component κ_N coinciding with $x_N + \Delta x_N$ which maximizes the increment of objective value.

The following properties of l are important:

1) The general constraints are maintained along direction l since

$$Ax(\theta) = Ax + \theta A(\kappa - x) = (1 - \theta)Ax + \theta A\kappa = b.$$

2) The objective value increases along l:

$$dF(x(\theta))/d\theta = c'l = -\Delta'_N l_N = \beta(x, J_{sup}) > \varepsilon \geq 0.$$

Owing to 2), the step length θ along l should be chosen without violating the simple constraints. This step is less than 1 because $x(1) = x + l = \kappa$, but the case $d_* \leq \kappa \leq d^*$ has not been realized.

For $0 \leq \theta < 1$ the non-support components $x_N(\theta)$ do not violate the simple constraints since $d_{*N} \leq x_N \leq d^*_N$ and $d_{*N} \leq \kappa_N \leq d^*_N$. Consequently increasing θ on the interval $[0,1[$ can violate only the simple support constraints

$$d_{*sup} \leq x_{sup}(\theta) \leq d^*_{sup}. \tag{4.3}$$

We find maximal θ^0 which provides fulfilment of (4.3). Write (4.3) in the component form

$$d_{*j} \leq x_j(\theta) = x_j + \theta l_j \leq d^*_j, \; j \in J_{sup}. \tag{4.4}$$

Denote by θ_j the maximal step length determined by the j-th constraint (4.4). The following three cases are possible for each j: 1) $l_j > 0$, 2) $l_j < 0$, 3) $l_j = 0$.

In Case 1) the component $x_j(\theta)$ increases and achieves the critical value d^*_j for $\theta = \theta_j = (d^*_j - x_j)/l_j$. Similarly in Case 2) the function $x_j(\theta)$ decreases and achieves the critical value d_{*j} for $\theta = \theta_j = (d_{*j} - x_j)/l_j$. In Case 3) the component $x_j(\theta)$ does not change $x_j(\theta) = x_j$, i.e. $\theta_j = \infty$.

Thus we obtain

$$\theta_j = \begin{cases} (d^*_j - x_j)/l_j & \text{for } l_j > 0, \\ (d_{*j} - x_j)/l_j & \text{for } l_j < 0, \\ \infty & \text{for } l_j = 0. \end{cases} \tag{4.5}$$

The maximal step length θ^0 with respect to the components $x_j(\theta)$, $j \in J_{sup}$, is equal to

$$\theta^o = \theta_{j_o} = \min \theta_j, \; j \in J_{sup}. \tag{4.6}$$

The index $j_o \in J_{sup}$ indicates the first component $x_{j_o}(\theta^o)$ which reaches the bound of simple constraints when θ is increased.

If $\{x, J_{sup}\}$ is a non-degenerate SF-point then $\theta^o > 0$ since $\theta_j > 0$, $j \in J_{sup}$.

Thus the feasible point $\bar{x} = x(\theta^o) = x + \theta^o l$ has been constructed. We calculate the suboptimality estimate for the SF-point $\{\bar{x}, J_{sup}\}$:

$$\beta(\bar{x}, J_{sup}) = \Delta'_N(\bar{x}_N - \kappa_N) = \Delta'_N(x_N + \theta^o l_N - \kappa_N) =$$

$$= \Delta'_N(1 - \theta^o)(x_N - \kappa_N) = (1 - \theta^o)\beta(x, J_{sup}) \le \beta(x, J_{sup}).$$

One can see that the transformation $x \longrightarrow \bar{x}$ satisfies the accepted principle: the suboptimality estimate does not increase and, in the case of non-degeneracy, strictly decreases.

If

$$\beta(\bar{x}, J_{sup}) = (1 - \theta^o)\beta(x, J_{sup}) \le \varepsilon$$

then the solution process is stopped: $\bar{x}$ is an ε-optimal feasible point. Otherwise we pass to the second part of the iteration.

4.2. Change of support. Short step rule .

Definition 4.1. The SF-point $\{x, J_{sup}\}$ is called dually non-degenerate if all the non-support components of the accompanying co-point are not equal to zero:

$$\delta_j = \Delta_j \neq 0, \ j \in J_N. \qquad (4.7)$$

The SF-point $\{x, J_{sup}\}$ having property

$$d_{*j} < x_j < d_j^*, \ j \in J_{sup} \qquad (4.8)$$

is called primally non-degenerate.

Further the SF-point will be called non-degenerate if it satisfies (4.7),(4.8).

As shown above, every support J_{sup} determines the accompanying dual feasible point λ which helps to calculate the degree of non-optimality of support:

$$\beta(J_{sup}) = \Phi(\lambda) - \Phi(\lambda^0).$$

Therefore we change the support as follows:

$$\begin{array}{ccc} J_{sup} & & \bar{J}_{sup} \\ \updownarrow & & \updownarrow \\ \lambda & \longleftrightarrow & \bar{\lambda} \end{array}, \qquad (4.9)$$

where a new dual feasible point $\bar{\lambda}$ satisfies the conditions i) $\Phi(\bar{\lambda}) \leq \Phi(\lambda)$, ii) $\bar{\lambda}$ is the accompanying dual feasible point for some support $\bar{J}_{sup}$. The degree of non-optimality does not increase:

$$\beta(\bar{J}_{sup}) = \Phi(\bar{\lambda}) - \Phi(\lambda^0) \leq \Phi(\lambda) - \Phi(\lambda^0) = \beta(J_{sup}),$$

for such a change.

To calculate $\bar{\lambda}$ we construct a curve $\lambda(\sigma) = (y(\sigma), v(\sigma), w(\sigma))$, $\sigma \geq 0$, in the space of dual feasible points. Since the coordinated dual feasible point $\lambda = (y, v, w)$ is fully determined by the first component y, we begin construction of the curve with the component $y(\sigma)$, $\sigma \geq 0$, which is projection of curve $\lambda(\sigma)$, $\sigma \geq 0$, in the space of variables y.

Determine a rule of variation of the support components

of the co-point:

$$\delta_{sup}(\sigma) = \delta_{sup} + \sigma\Delta\delta_{sup}, \tag{4.10}$$

where

$$\Delta\delta_{sup} = -e_{j_o} sign\ l_{j_o},\ l_{j_o} = \kappa_{j_o} - x_{j_o}, \tag{4.11}$$

and the index j_o is found by calculating the step length θ^o in Section 4.1. According to (4.10),(4.11), among the support components of the co-point we change only δ_j, corresponding to the component of the feasible point $\bar{x}_{j_o} = x_{j_o}(\theta^o)$ which come first to the bound. The component δ_{j_o} is changed in such a way that its new value $\delta_{j_o}(\sigma)$ satisfies (2.10) for $\sigma\geq 0$ together with the component $\bar{x}_{j_o}$. Owing to (4.10), (4.11) the last property holds true because

$$\begin{aligned} &\text{for } l_{j_o} < 0 \text{ we have } \delta_{j_o}(\sigma) \geq \delta_{j_o} = 0,\ \bar{x}_{j_o} = d_{*j_o}, \\ &\text{for } l_{j_o} > 0 \text{ we have } \delta_{j_o}(\sigma) \leq \delta_{j_o} = 0,\ \bar{x}_{j_o} = d^*_{j_o}. \end{aligned} \tag{4.12}$$

For given $\delta_{sup}(\sigma)$, $\sigma \geq 0$, and the equation

$$\delta'_{sup} = y'(\sigma)P - c'_{sup}$$

we find the projection $y(\sigma)$, $\sigma\geq 0$ of the desired curve

$$y'(\sigma) = (\delta_{sup}(\sigma) + c_{sup})'Q = (\delta_{sup} + c_{sup})'Q +$$

$$= \sigma\Delta\delta'_{sup}Q = y' + \sigma\Delta y, \tag{4.13}$$

where $y' = c'_{sup}Q$, $\Delta y' = \Delta\delta'_{sup}Q = -e'_{j_o}Q\ sign\ l_{j_o} = -q'_{j_o}\ sign\ l_{j_o}$.

Knowing $y(\sigma)$ we calculate the non-support component of the coordinated co-point

$$\delta'_N(\sigma) = y'(\sigma)A_N - c'_N = y'A_N - c'_N + \sigma\Delta y'A_N = \delta'_N + \sigma\Delta\delta'_N, \quad (4.14)$$

where

$$\Delta\delta'_N = \Delta y'A_N = -q'_{j_o}A_N sign\ l_{j_o}. \quad (4.15)$$

The projections $v(\sigma)$, $w(\sigma)$, $\sigma \geq 0$ of the curve $\lambda(\sigma), \sigma \geq 0$ are constructed according to (3.16) and (4.10), (4.11), (4.14), (4.15).

We consider at first the case when $\{x, J_{sup}\}$ is a dually non-degenerate SF-point. In this case according to (4.14) one can find an interval $0 \leq \sigma \leq \sigma^*, \sigma^* > 0$ on which signs of values $\delta_j(\sigma)$ coincide with signs of δ_j for all $j \in J_N$. Therefore

$$v_j(\sigma) = \delta_j(\sigma) = \delta_j + \sigma\Delta\delta_j,\ w_j(\sigma) = 0, \quad \text{if} \quad \delta_j > 0,$$

$$v_j(\sigma) = 0,\ w_j(\sigma) = -\delta_j(\sigma) = -\delta_j - \sigma\Delta\delta_j, \quad \text{if} \quad \delta_j < 0,\ j \in J_N;$$

$$v_j(\sigma) = w_j(\sigma) = 0,\ j \in J_{sup} \setminus j_o; \quad (4.16)$$

$$v_{j_o}(\sigma) = \delta_{j_o}(\sigma) = \sigma,\ w_{j_o}(\sigma) = 0 \text{ for } \Delta\delta_{j_o} = 1 \text{ (i.e. } l_{j_o} < 0),$$

$$v_{j_o}(\sigma) = 0,\ w_{j_o}(\sigma) = -\delta_{j_o}(\sigma) = -\sigma \text{ for } \Delta\delta_{j_o} = -1 \text{ (i.e. } l_{j_o} > 0).$$

Relations (4.10)-(4.16) describe the curve $\lambda(\sigma)$ in the region where the signs of $\delta_j(\sigma)$, $j \in J_N$, are not changed.

We explain the behaviour of the dual objective function $\Phi(\lambda)$ in the constructed region of the curve $\lambda(\sigma)$, $\sigma \geq 0$.

Because of (4.10), (4.16) we have

$$\begin{aligned}\Phi(\lambda(\sigma)) &= b'y(\sigma) - d'_*v(\sigma) + d^{*\prime}w(\sigma) = b'y(\sigma) - \kappa'_N\delta_N(\sigma) - \\ &-d_{*j_o}v_{j_o}(\sigma) + d^*_{j_o}w_{j_o}(\sigma) = b'y + \sigma b'\Delta y - \kappa'_N\delta_N - \sigma\kappa'_N\Delta\delta_N - \\ &- \sigma\bar{x}_{j_o}\Delta\delta_{j_o} = \Phi(\lambda) + \sigma(b'\Delta y - \kappa'_N\Delta\delta_N - \bar{x}_{j_o}\Delta\delta_{j_o}) =\end{aligned} \quad (4.17)$$

$$= \Phi(\lambda) + \sigma(\kappa'_{sup}\Delta\delta_{sup} - \bar{x}_{j_o}\Delta\delta_{j_o}) = \Phi(\lambda) - \sigma|\kappa_{j_o} - \bar{x}_{j_o}|.$$

Thus on the initial arc of the curve $\lambda(\sigma)$, $\sigma\geq 0$, the dual objective function is linear with respect to σ and decreases at a constant rate:

$$\alpha = -|\kappa_{j_o} - \bar{x}_{j_o}| < 0. \tag{4.18}$$

According to the above calculations a linear rule of variation of function $\Phi(\lambda(\sigma))$, $\sigma \geq 0$ remains valid until the first zero of the non-support components $\delta_j(\sigma)$, $j \in J_N$ appears. The value σ^1 determined by this zero can easily be calculated from (4.14):

$$\sigma^1 = \sigma_{j_1} = \min_{j\in J_N} \sigma_j,$$

$$\sigma_j = \begin{cases} -\delta_j/\Delta\delta_j, & \text{if } \delta_j\Delta\delta_j < 0; \\ \infty, & \text{otherwise.} \end{cases} \tag{4.19}$$

We note here that in the case of dual degeneracy the components δ_j, $j \in J_N$ can take zero values. In Section 3 we agree to distinguish $\delta_j = +0$ and $\delta_j = -0$ by the value d_{*j} or d^*_j of the j-th component of the pseudo-feasible point κ. This distinction should be taken into account in (4.19). In particular we consider that

$$\delta_j\Delta\delta_j \begin{cases} <0, & \text{if } \delta_j = +0,\ \Delta\delta_j < 0 \text{ or } \delta_j = -0,\ \Delta\delta_j > 0, \\ >0, & \text{if } \delta_j = +0,\ \Delta\delta_j > 0 \text{ or } \delta_j = -0,\ \Delta\delta_j < 0. \end{cases}$$

Set

$$\sigma^* = \sigma^1 = \sigma_{j_1},\ \bar{\lambda} = \lambda(\sigma^*),\ \bar{J}_{sup} = (J_{sup} \setminus j_o) \cup j_1.$$

By construction we have

$$\Phi(\bar{\lambda}) = \Phi(\lambda) + \sigma^*\alpha < \Phi(\lambda). \tag{4.20}$$

Show that $\bar{J}_{sup}$ is a support, and that $\bar{\lambda}$ is an accompanying dual feasible point.

Among the numbers σ_j, $j \in J_N$ (4.19) one can always find finite ones, since otherwise $\Phi(\lambda(\sigma)) \longrightarrow -\infty$ for $\sigma \longrightarrow \infty$ but that is impossible, owing due to (3.12) and the consistency of constraints of the primal problem.

According to (4.19) the finiteness of σ_j implies that $\Delta\delta_{j_1} \neq 0$. The number $\mu = |\Delta\delta_{j_1}|$ has the representation

$$\mu = |q'_{j_o} a_{j_1}| = |e'_{j_o} Q a_{j_1}|.$$

The matrix $\bar{P} = A(I, \bar{J}_{sup})$ obtained from $P = A(I, J_{sup})$ by exchanging the column a_{j_o} for a_{j_1} is non-singular if $\mu \neq 0$ and the matrix P is non-singular. Consequently $\bar{J}$ is a support.

By construction $\bar{\lambda}$ is a coordinated dual feasible point. For the co-point δ we get $\bar{\delta}_{sup} = \bar{\delta}(\bar{J}_{sup}) = 0$:

$$\bar{\delta}_{j_1} = 0,\ \bar{\delta}(\bar{J}_{sup}\backslash j_1) = \bar{\delta}(J_{sup}\backslash j_o) = \delta(J_{sup}\backslash j_o) = 0$$

i.e. $\bar{\lambda}$ is a dual feasible point accompanying the support $\bar{J}_{sup}$.

Thus the scheme (4.9) has been completely realized.

The described procedure of support variation $J_{sup} \longrightarrow \bar{J}_{sup}$ will be called a short step rule. According to (4.20) the degree of the support non-optimality decreases by the value $|\alpha|\sigma^*$ at the exchange. This value is positive if the SF-point $\{x, J_{sup}\}$ is dually non-degenerate.

Because of (3.22),(3.23) the suboptimality estimate decreases by the same value

$$\beta(\bar{x}, \bar{J}_{sup}) = \beta(\bar{x}, J_{sup}) + \alpha\sigma^* < \beta(\bar{x}, J_{sup}). \tag{4.21}$$

The transition $J_{sup} \longrightarrow \bar{J}_{sup}$ completes the second part of the iteration. The method for solving problem (1.2), in which the short step rule is used, is called an adaptive method with short step. Upon iteration of the method we have converse of the suboptimality estimate:

$$\beta(\bar{x},\bar{J}_{sup}) = (1 - \theta^{0})\beta(x,J_{sup}) + \alpha\sigma^{*}.$$

It strictly decreases if the SF-point $\{x,J_{sup}\}$ is primally or dually non-degenerate.

Let $\beta(\bar{x},\bar{J}_{sup}) \le \varepsilon$. Then $\bar{x}$ is an ε-optimal feasible point. Otherwise we pass to a new iteration with the SF-point $\{\bar{x},\bar{J}_{sup}\}$.

4.3. Change of support. Long step rule.

In the previous section the support was changed under relaxation of the dual objective function along the curve $\lambda(\sigma)$, $\sigma \ge 0$ at the moment when the first zero of the non-support components $\delta_j(\sigma)$, $j \in J_N$ appeared. This property provided the name short step rule. Below we consider the possibility of further movement along $\lambda(\sigma)$, $\sigma \ge 0$.

We explain the behaviour of the function $\Phi(\lambda(\sigma))$, $\sigma \ge 0$ along all $\lambda(\sigma)$, $\sigma \ge 0$. Assume at the beginning that $\{x,J_{sup}\}$ is the dually non-degenerate SF-point and at $\sigma = \sigma^{1}$ the only component of $\delta_N(\sigma)$ takes the value zero:

$$\delta_{j_1}(\sigma^{1}) = 0,\ \delta_j(\sigma^{1}) \ne 0, j \in J_N \setminus j_1 .$$

The linearity of the dual objective function

$$\Phi(\lambda(\sigma)) = b'y(\sigma) - d_{*}'v(\sigma) + d^{*\prime}w(\sigma)$$

is violated at $\sigma=\sigma^{1}=\sigma_{j_1}$ only for one expression

$$\xi_{j_1}(\sigma) = -d_{*j_1}v_{j_1}(\sigma) + d^{*}_{j_1}w_{j_1}(\sigma). \tag{4.22}$$

The investigation shows that the rate of change of (4.22) at $\sigma = \sigma^{1}$ has a positive increment

$$\Delta\alpha^{\xi}_{j_1} = (d^{*}_{j_1} - d_{*j_1})|\Delta\delta_{j_1}|. \tag{4.23}$$

For the dually non-degenerate SF-point $\{x, J_{sup}\}$, let several components of the function $\delta_N(\sigma)$, $\sigma \ge 0$ become zero simultaneously at $\sigma = \sigma^1$ and let σ^{k-1}, σ^k, $0 < \sigma^{k-1} < \sigma^k$, be two arbitrary neighbouring zeros of the functions $\delta_j(\sigma)$, $\delta \ge 0$, $j \in J_N$: α^{k-1} be the rate of dual objective function change on $\sigma^{k-1} \le \sigma \le \sigma^k$; $J_k = \{ j \in J_N : \delta_j(\sigma^k) = 0\}$. We get

$$\Delta\alpha^k = \sum_{j \in J_k} (d_j^* - d_{*j}) |\Delta\delta_j| \tag{4.24}$$

and

$$\alpha^k = \alpha^{k-1} + \Delta\alpha^k \tag{4.25}$$

on $\sigma^k \le \sigma \le \sigma^{k+1}$.

It is shown in Section 3 that the degree of non-optimality of the support decreases together with the dual objective function. Therefore we find the point $\sigma = \sigma^*$ at which the function $\Phi(\lambda(\sigma))$, $\sigma \ge 0$, achieves the minimal value along $\lambda(\sigma), \sigma \ge 0$. It can easily be done because the rule of rate variation has already been found: that is (4.24), (4.25). The function $\Phi(\lambda(\sigma))$, $\sigma \ge 0$ achieves the minimum at $\sigma = \sigma^* = \sigma^{k_o}$,

$$\alpha^{k_o - 1} < 0, \ \alpha^{k_o} \ge 0. \tag{4.26}$$

The index k_o exists always since $\inf \Phi(\lambda(\sigma)) > -\infty$. Set

$$\bar{\lambda} = \lambda(\sigma^*). \tag{4.27}$$

By construction we have

$$\Phi(\bar{\lambda}) = \Phi(\lambda) - \sum_{k=o}^{k_o - 1} \alpha^k (\sigma^{k+1} - \sigma^k), \ \sigma^o = 0. \tag{4.28}$$

To complete the realization of the scheme (4.9) we should construct a new support $\bar{J}_{sup}$ and show that $\bar{\lambda}$ is an accompanying dual feasible point. A new support will be constructed in the form

$$\bar{J}_{sup} = (J_{sup} \setminus j_o) \cup j_*.$$

To find a suitable index j_* we analyse two possibilities:

1) $|J_{k_o}| = 1$; 2) $|J_{k_o}| > 1$.

In Case 1) the set J_{k_o} consists of the only element which is taken as j_*.

In Case 2) we arrange elements of the set J_k in an arbitrary way (for example, in increasing value):

$$s_1, s_2, \dots, s_q; \quad \bigcup_{i=1}^{q} s_i = J_{k_o}; \; s_1 < s_2 < \dots < s_q. \tag{4.29}$$

Sorting out indices from J_{k_o} in the order (4.29) we find s_p for which

$$\alpha_{s_{p-1}} = \alpha^{k_o+1} + \sum_{i=1}^{p-1} (d^*_{s_i} - d_{*s_i})|\Delta\delta_{s_i}| < 0,$$

$$\alpha_{s_p} = \alpha_{s_{p-1}} + (d^*_{s_p} - d_{*s_p})\Delta\delta_{s_p} \geq 0. \tag{4.30}$$

It follows from (4.26) that s_p exists. We set $j_* = s_p$. The set $\bar{J}_{\text{sup}}$ has been constructed.

Just as in Section 2, it can be shown that $\bar{J}_{\text{sup}}$ is a support and $\bar{\lambda}$ is an accompanying dual feasible point. The scheme (4.9) has been completely realized.

This procedure is called the long step rule. For $k_o = 1$ it becomes the short step rule.

The transition $J_{\text{sup}} \longrightarrow \bar{J}_{\text{sup}}$ completes the second part of the iteration $\{x, J_{\text{sup}}\} \longrightarrow \{\bar{x}, \bar{J}_{\text{sup}}\}$ of the adaptive method with long dual step length in which the first part (transformation $x \longrightarrow \bar{x}$) is carried out by the scheme of Section 4.1.

According to (4.25),(3.23) the suboptimality estimate of the SF-point $\{x, J_{\text{sup}}\}$ is equal to

$$\beta(\bar{x}, \bar{J}_{\text{sup}}) = (1 - \theta^o)\beta(x, J_{\text{sup}}) - \sum_{k=o}^{k_o-1} \alpha^k(\sigma^{k+1} - \sigma^k).$$

If $\beta(\bar{x}, \bar{J}_{\text{sup}}) \leq \varepsilon$ then $\bar{x}$ is an ε-optimal feasible point of problem (1.2). Otherwise we pass to a new iteration with the SF-point $\{\bar{x}, \bar{J}_{sup}\}$.

To conclude the subsection we shall show modifications which should be introduced into the rule of support change where $\{x, J_{sup}\}$ is a dually degenerate SF-point. For construction of an accompanying pseudo-feasible point we agreed that the set

$$\{ j \in J_N : \Delta_j = 0 \}$$

is divided into two parts J_N^+ and J_N^-. The numbers $\Delta_j = 0$, $j \in J_N^+$ were assumed to be infinitesimally positive and $\kappa_j = d_{*j}$, $j \in J_N^+$ were set. The numbers $\Delta_j = 0$, $j \in J_N^-$ were assumed to be infinitesimally negative and $\kappa_j = d_j^*$, $j \in J_N^-$, were set.

As shown above, every transition over zero of the functions $\delta_j(\sigma)$, $j \in J_N$ causes a jump in the rate of dual objective function variation. Similar jumps are caused by transitions of the functions $\delta_j(\sigma)$, $j \in J_N^+$ $(\Delta\delta_j < 0)$ and $j \in J_N^-$ $(\Delta\delta_j > 0)$ over zero at the point $\sigma = 0$:

$$\Delta\alpha^0 = \sum_{j \in J_0} (d_j^* - d_{*j}) |\Delta\delta_j| > 0$$

where $J_0 = \{j \in J_N^+, \Delta\delta_j < 0 \text{ or } j \in J_N^-, \Delta\delta_j > 0\}$.

Therefore in the case of dual degeneracy the rate of dual objective function variation on the initial interval $0 = \sigma^0 \leq \sigma < \sigma^1$ is equal to

$$\alpha^0 = -|\kappa_{j_0} - \bar{x}_{j_0}| + \sum_{j \in J_0} (d_j^* - d_{*j}) |\Delta\delta_j| .$$

The rate α^0 can be positive for $J_0 \neq \emptyset$. Therefore the step length $\sigma^* = \sigma^0 = 0$ is possible.

4.4. Discussion.

The method is called adaptive owing to its property of using all the initial and current information for effective construction of suboptimal feasible points. The adaptive method belongs to the same class as the primal simplex method [11]. However, the simplex method uses not arbitrary points but special basic points, of which all the non-support (non-basic) components are critical. The only non-support component of the feasible point is changed upon iterations of the simplex method. The support (basis of the simplex method) is changed together with the feasible point and its degree of non-optimality can increase upon iterations. To stop the solution process the simplex method uses (in the case of existence of solution) only the optimality criterion since it has no suboptimality criterion at all.

Various modifications of the adaptive method are considered in [16,19].

4.5. Algorithm.

We describe the algorithm of the adaptive method with long steps.

We start solving the problem assuming the initial SF-point $\{x, J_{sup}\}$, the matrix $Q = A(I, J_{sup})^{-1}$ and the method parameters $\varepsilon \geq 0$, $\mu \geq 0$ to be known.

Step 1. Calculate the multipliers and the support gradient:

$$u' = c'_{sup} Q, \qquad \Delta = u'A - c,$$

and divide the set J_N into two non-intersecting parts

$$J_N^+ = \{ j \in J_N : \Delta_j \geq 0 \}, \quad J_N^- = \{ j \in J_N : \Delta_j < 0 \},$$

$$J_N^+ \cup J_N^- = J_N, \quad J_N^+ \cap J_N^- = \varnothing.$$

Determine non-support components of the accompanying pseudo-feasible point κ_N:

$$\kappa_j = \begin{cases} d_{*j} & \text{for } j \in J_N^+, \\ d_j^* & \text{for } j \in J_N^-. \end{cases}$$

Calculate the suboptimality estimate

$$\beta(x,J_{sup}) = \sum_{j\in J_N} \Delta_j(x_j-\kappa_j).$$

If $\beta(x,J_{sup}) \le \varepsilon$, then STOP, x is an ε-optimal feasible point.

Step 2. Calculate the support components of the pseudo-feasible point

$$\kappa_{sup} = Q(b - A_N\kappa_N).$$

If $d_{*sup} \le \kappa_{sup} \le d^*_{sup}$, then STOP, κ is an optimal feasible point. For

$$d_{*j} - \mu \le \kappa_j \le d^*_j + \mu, \ \forall \ j \in J_{sup}$$

we begin the support change (go to *Step* 3) choosing any $j \in J_{sup}$ for which $\kappa_j \notin [d_{*j}, d^*_j]$ as j_o.

Otherwise determine

$$l = \kappa - x$$

and calculate the step θ^o along l:

$$\theta = \min_{j\in J_{sup}} \theta_j = \theta_{j_o}, \ j_o \in J_{sup}$$

where

$$\theta_j = \begin{cases} (d^*_j-x_j)/l_j & \text{for } l_j>0, \\ (d_{*j}-x_j)/l_j & \text{for } l_j<0, \\ \infty & \text{for } l_j=0, \ j\in J_{sup}. \end{cases}$$

Change the feasible point x

$$\bar{x} = x + \theta^o l.$$

Calculate

$$\beta(\bar{x},J_{sup})=\beta(x,J_{sup})(1-\theta^o).$$

If $\beta(\bar{x}, J_{sup}) \le \varepsilon$, then STOP, $\bar{x}$ is an ε-optimal feasible point. Otherwise goto *Step* 3.

Step 3. Construct the direction $\Delta\delta_N$ for changing the non-support component of the co-point $\delta=\Delta$:

$$\Delta\delta'_N = \nu q'_{j_o} A_N$$

where

$$\nu = \begin{cases} +1 & \text{for } \bar{x}_{j_o} = d_{*j_o}, \\ -1 & \text{for } \bar{x}_{j_o} = d^*_{j_o}. \end{cases}$$

For every $j \in J_N$ we calculate such σ_j that $\delta_j(\sigma) =$ $= \delta_j + \sigma\Delta\delta_j = 0$. We get

$$\sigma_j = \begin{cases} -\Delta_j/\Delta\delta_j, & \text{if } \Delta_j\Delta\delta_j < 0 \text{ or } j \in J_N^+,\ \Delta\delta_j < 0;\ \text{or } j \in J_N^-,\ \Delta\delta_j > 0; \\ \infty & \text{in other cases.} \end{cases}$$

Step 4. Find j_* to be added to the support. Arrange the indexes $\{j \in J_N : \sigma_j \ne \infty\}$ in increasing values σ_j:

$$\sigma_{j_1} \le \sigma_{j_2} \le \ldots \le \sigma_{j_p};\quad j_k \in J_N,\quad \sigma_{j_k} \ne \infty,\ k = \overline{1,p}.$$

For every j_k, $k = \overline{1,p}$ we calculate the jump of the rate of the dual objective function

$$\Delta\alpha_{j_k} = |\Delta\delta_{j_k}|(d^*_{j_k} - d_{*j_k}|.$$

As j_* we choose j_q such that

$$\alpha_{j_{q-1}} = \alpha_o + \sum_{k=1}^{q-1} \Delta\alpha_{j_k} < 0,\quad \alpha_{j_q} = \alpha_o + \sum_{k=1}^{q} \Delta\alpha_{j_k} \ge 0$$

where $\alpha_o = -|\kappa_{j_o} - \bar{x}_{j_o}|$ is the initial rate of dual objective function change.

For each $k = \overline{1,q-1}$ we set

$$\bar{\kappa}_j = \begin{cases} d_{*j} & \text{for } \Delta\delta_{j_*} > 0, \\ d_j^* & \text{for } \Delta\delta_{j_*} < 0, \end{cases}$$

adding simultaneously j_k with $\Delta\delta_j > 0$ to $\bar{J}_N^+$ and j_k with $\Delta\delta_j < 0$ to $\bar{J}_N^-$.

We calculate

$$\Delta\Phi = \Phi(\lambda(\sigma_{j_q})) - \Phi(\lambda) = \sum_{k=1}^{q} \alpha_{j_{k-1}} (\sigma_{j_k} - \sigma_{j_{k-1}})$$

where $\alpha_{j_o} = \alpha_o$, $\sigma_{j_o} = \sigma_o = 0$.

If

$$\beta(\bar{x},\bar{J}_{\text{sup}}) = \beta(\bar{x},J_{\text{sup}}) + \Delta\Phi \le \varepsilon$$

then STOP, $\bar{x}$ is an ε-optimal feasible point. Otherwise modify the support

$$J_{\text{sup}} \longrightarrow \bar{J}_{\text{sup}} = (J_{\text{sup}} \backslash j_o) \cup j_*$$

and pass to a new iteration with the SF-point $\{\bar{x},\bar{J}_{\text{sup}}\}$ and the sets $\bar{J}_N^+$, $\bar{J}_N^-$.

5. PHASE I.

5.1. Construction of an initial feasible point .

In designing new productions the results of solving the simplified problems can be used for composing initial feasible points. In this case some vector $\tilde{x}$ is known along with the mathematical model. This vector, as a rule, is not a feasible point but in a certain sense it is close to the optimal one. Therefore it is natural to use the vector $\tilde{x}$ for constructing an initial feasible point. Since the simple constraints can be easily satisfied we assume that the vector $\tilde{x}$ satisfies the simple constraints $d_* \le \tilde{x} \le d^*$ but

$$\omega = b - A\tilde{x} \ne 0.$$

We introduce an auxiliary LP problem which we call the Phase I problem:

$$- \sum_{i=1}^{m} x_{n+i} \longrightarrow max,$$

$$A(i,J)x + a_{i,n+i}x_{n+i} = b_i, \; i = \overline{1,m}; \qquad (5.1)$$

$$d_* \le x \le d^*, \; 0 \le x_{n+i} \le |\omega_i|, \; i = \overline{1,m},$$

where $a_{i,n+i} = 1$ for $\omega_i \ge 0$; $a_{i,n+i} = -1$ for $\omega_i < 0$, $i=\overline{1,m}$.

Problem (5.1) has $n+m$ variables. We denote their indices by $K = J \cup J^a$, $J^a = \{n+1, n+2, \ldots, n+m\}$. The variables x_j, $j \in J^a$ are called artificial. These variables caracterize a degree of the i-th general constraint violation. It is obvious that the vector $z = (x = \tilde{x}, \; x_{n+i} = |\omega_i|, \; i = \overline{1,m})$ is a feasible point of problem (5.1). As an initial support we take the set $K_{sup} = J^a = \{ n+1, \ldots, n+m \}$. Problem (5.1) always has a solution since the objective function is bounded above. We solve it using the adaptive method. Let

$\{z^*, K^*_{sup}\}$, $z^* = (x^*, x^*_{n+i}, i = \overline{1,m})$ be a solution of problem (5.1).

Analyse the following possibilities:

1) $\sum_{i=1}^{m} x^*_{n+i} > 0$,

2) $\sum_{i=1}^{m} x^*_{n+i} = 0, \quad K^*_{sup} \subset J$,

3) $\sum_{i=1}^{m} x^*_{n+i} = 0, \quad K^*_{sup} \cap J^a \neq \emptyset$.

In Case 1) the solution process for the initial problem is stopped because

its constraints are inconsistent.

In Case 2) the pair $\{x^*, J_{sup}\}$, $J_{sup} = K^*_{sup}$ is an SF-point of the original problem.

Examine Case 3). Just as in Case 2) the vector x^* is a feasible point of problem (1.2). We find the support J_{sup} deleting the indexes $n + i \in J^a$ of artificial variables from the set K^*_{sup}. Let $I^* = \{ i \in I: n+i \in K^*_{sup}\}$. Introduce an auxiliary problem

$$c'x \longrightarrow max,$$

$$A(i,J)x = b_i, \; i \in I\backslash I^*; \qquad (5.2)$$

$$A(i,J)x+a_{i,n+i}x_{n+i} = b_i, \; i \in I^*;$$

$$d_* \le x \le d^*, \; d_{*n+i} \le x_i \le d^*_{n+i}, \quad d_{*n+i} = d^*_{n+i} = 0, \; i \in I^*.$$

Since $x_{n+i} = 0$, $i \in I^*$, problem (5.2) is equivalent to (1.2). The vector $z^* = (x^*, x^*_{n+i} = 0, i \in I^*)$ is its feasible point and K^*_{sup} is a support. The procedure of "clearing" the set K^*_{sup} and constructing the initial support J_{sup} of the primal problem is described in [16].

Upon each iteration the artificial index $i \in I^*$ is deleted from problem (5.2). The process is completed after

$|I^*|$ iterations with the support $\hat{K}_{sup} \subset J$ of problem (1.2) without artificial indices and without linearly dependent rows in the general constraints.

5.2. Construction of an initial support.

We consider the general case of constructing an initial support. Assume that the initial feasible point and some elements of the support are known along with the mathematical model. Introduce some necessary notions.

Definition 5.1. The collection $M_{sup} = \{I_{sup}, J_{sup}\}$, $I_{sup} \subset I$, $J_{sup} \subset J$, $|I_{sup}| = |J_{sup}|$, is called a presupport if $\det A(I_{sup},J_{sup}) \neq 0$. A pair $\{x,M_{sup}\}$ will be called a presupport feasible (PSF-) point. To simplify notation we shall set $P = P(I_{sup},J_{sup}) = A(I_{sup},J_{sup})$, $Q = Q(J_{sup},I_{sup}) = P^{-1}$. Construct the co-point accompanying the presupport

$$\Delta' = u'A - c'$$

where

$u(I) = (u_{sup},u_N)$, $u'_{sup} = u'(I_{sup}) = c'_{sup}Q$, $u_N = u(J_N) = 0$, $c_{sup}=c(J_{sup})$.

Divide the set $J_N = J \backslash J_{sup}$ into two non-intersecting subsets :

$J_N^+ = \{ j \in J_N : \Delta_j \geq 0\}$, $J_N^- = \{ j \in J_N : \Delta_j \leq 0 \}$, $J_N = J_N^+ \cup J_N^-$.

Introduce the pseudo-feasible point κ accompanying the pre-support:

$$\kappa_j = d_{*j},\ j \in J_N^+,\ \kappa_j = d_j^*,\ j \in J_N^-,$$

$$\kappa_{sup} = Q(b(I_{sup}) - A(I_{sup},J_N)\kappa_N).$$

The vector $\lambda = (u,v,w)'$ with coordinated components v, w satisfies the constraints of the dual problem (3.14). It is a dual feasible point accompanying the presupport. The value

$$\beta(x,M_{sup}) = \Delta'_N(x_N - \kappa_N)$$

is called a suboptimality estimate of the PSF-point since

$$F(x^0) - F(x) \le \beta(x,M_{sup}).$$

The suboptimality estimate admits the decomposition

$$\beta(x,M_{sup}) = \beta(x) + \beta(M_{sup})$$

where $\beta(x) = F(x^0) - F(x)$ is a degree of feasible point non-optimality, $\beta(M_{sup}) = \Phi(\lambda) - \Phi(\lambda^0)$ is a degree of pre-support non-optimality calculated with help of the dual objective function at the optimal λ^0 and accompanying λ dual feasible points. For any $\varepsilon \ge 0$ the feasible point x is ε-optimal if and only if there exists a presupport M_{sup} such that the inequality $\beta(x,M_{sup}) \le \varepsilon$ is fulfilled at the PSF-point $\{x,M_{sup}\}$.

All the above statements follow from the results of the interval problem presented in Section 4. On the basis of these we proceed to constructing the initial support.

Take an arbitrary presupport $_{sup} = \{I_{sup}, J_{sup}\}$, $|I_{sup}| \le rank\ A$. The empty set $M_{sup} = \emptyset$ with $u_{sup} = 0$, $\kappa = \kappa_N$ can be taken as M_{sup}. Calculate for M_{sup} the value $\beta(x,M_{sup}) = \Delta'_N(x_N - \kappa_N)$. If $\beta(x,M_{sup}) \le \varepsilon$ then constructing the initial support is unnecessary since x is the ε-optimal feasible point. The solution process for (1.2) is stopped.

For $\beta(x,M_{sup}) > \varepsilon$ we construct the support component of the accompanying pseudo-feasible point:

$$\kappa_{sup} = Q(b(I_{sup}) - A(I_{sup},J_N)\kappa_N).$$

If $A\kappa = b$, $d_{*sup} \le \kappa_{sup} \le d^*_{sup}$ then κ is the optimal feasible point and it is unnecessary to construct the initial support. Problem (1.2) has been solved.

Let $A\kappa = b$ but the inequalities $d_{*sup} \le \kappa_{sup} \le d^*_{sup}$ be violated. Then according to the adaptive method we construct a new PSF-point $\{\bar{x}, \bar{M}_{sup}\}$, $\bar{x} = x + \theta^0 l$, $l = \kappa - x$;

$\bar{M}_{sup} = \{ I_{sup}, \bar{J}_{sup} \}$, $\bar{J}_{sup} = (J_{sup} \setminus j_o) \cup j_*$, and go on searching the initial support.

Consider the case $\omega = b - A\kappa \neq 0$. By construction $\omega_{sup} = \omega(I_{sup}) = 0$. Nonzero residuals $\omega_i \neq 0$, $i \in I$ can appear only for $i \in I_N$. Denote by i_o the index such that

$$|\omega_{i_o}| = |b_{i_o} - A(i_o, J)\kappa| = \max_{i \in I_N} |b_i - A(i,J)\kappa|.$$

To transform the pre-support M_{sup} we construct the curve $\lambda(\sigma) = (y(\sigma), v(\sigma), w(\sigma))$, $\sigma \geq 0$, using coordinated dual feasible points. Begin with the vector $y(\sigma)$:

$$y_{i_o}(\sigma) = y_{i_o} + \sigma\Delta y_{i_o} = y_{i_o} - \sigma \text{ sign } \omega_{i_o}, \quad y_{i_o} = 0,$$

$$y_{sup}(\sigma) = y(I_{sup}) + \sigma\Delta y(I_{sup}),$$

$$\Delta y'(I_{sup}) = \Delta y'_{i_o} A(i_o, J_{sup})Q, \qquad (5.3)$$

$$y_i(\sigma) = 0, \; i \in I_N \setminus i_o.$$

According to (5.3) the co-point will vary as follows:

$$\delta'_N(\sigma) = y'(\sigma)A_N - c'_N = y'_{sup}(\sigma)A(I_{sup}, J_n) + y_{i_o}(\sigma)A(i_o, J_n) - c'_N =$$

$$= \delta'_N + \sigma\Delta\delta'_N, \quad \delta_{sup}(\sigma) \equiv \delta_{sup},$$

where $\delta_N = \Delta_N$, $\Delta\delta_N = A(i_o, J_N) \text{ sign } \omega_{i_o}$.

If $\Delta_j \neq 0$, $j \in J_N$ then the signs of $\delta_j(\sigma)$, $j \in J_N$ coincide with these of δ_j, $j \in J_N$ on $0 \leq \sigma \leq \sigma^*$, $\sigma^* > 0$. Therefore

$$v_j(\sigma) = \delta_j, \; w_j(\sigma) = 0, \; j \in J^+_N; \qquad (5.4)$$

$$v_j(\sigma) = 0, \; w_j(\sigma) = -\delta_j, \; j \in J^-_N; \; 0 \leq \sigma \leq \sigma^*.$$

The formulae (5.3),(5.4) determine the curve $\lambda(\sigma)$, $\sigma \geq 0$ on

$0 \le \sigma \le \sigma^*$. Calculate the dual objective value $\Phi(\lambda)$ on the initial region of the curve $\lambda(\sigma)$, $\sigma \ge 0$:

$$\Phi(\lambda(\sigma)) = b'y(\sigma) - d_*'v(\sigma) + d^{*}{}'w(\sigma). \qquad (5.5)$$

According to (5.4), $\Phi(\lambda(\sigma))$, $\sigma \ge 0$, is linear on $0 \le \sigma \le \sigma^*$ and decreases with the rate

$$\alpha_o = -|\omega_{i_o}|.$$

Just as in Section 4 one can show that in the case when zero values are present among δ_j, $j \in J_N$ the initial rate of decrease of $\Phi(\lambda(\sigma))$, $\sigma \ge 0$ is equal to

$$\alpha^o = -|\omega_{i_o}| + \sum_{j \in J_o} |\Delta\delta_j|(d_j^* - d_{*j}),$$

where $J_o = \{j \in J_N^+,\ \delta_j = 0,\ \Delta\delta_j < 0\} \cup \{j \in J_N^-,\ \delta_j = 0,\ \Delta\delta_j > 0\}$.

We introduce the following notations: $0 < \sigma^1 < \sigma^2 < \dots < \sigma^k$ are zeros of functions $\delta_j(\sigma)$, $j \in J_N$; $J_k = \{j \in J_N:\ \delta_j(\sigma^k) = 0\}$, α^k is the variation rate of function $\Phi(\lambda(\sigma))$, $\sigma \ge 0$, on $\sigma^k \le \sigma \le \sigma^{k+1}$. By analogy with Section 4 one can show that $\Phi(\lambda(\sigma))$, $\sigma \ge 0$ is a convex piecewise-linear function, the derivative of which on $\sigma^k \le \sigma \le \sigma^{k+1}$ is equal to

$$\alpha^k = \sigma^{k-1} + \Delta\alpha^k, \quad \Delta\alpha^k = \sum_{j \in J_k} |\Delta\delta_j|(d_j^* - d_{*j}).$$

Let k_o be an index such that

$$\alpha^{k_o} < 0, \quad \alpha^{k_o+1} \ge 0.$$

Then the dual objective function achieves the minimum at the point $\sigma^* = \sigma^{k_o}$. We order the set J_{k_o} of: $s_1, s_2, \dots, s_q$ and find among them the first index s_p for which the inequality

$$\alpha^{k_o} + \sum_{k=1}^{p} |\Delta\delta_{s_k}|(d_{s_k}^* - d_{*s_k}) \ge 0$$

holds.

Set $j_*=s_p$ and construct the new presupport

$$\bar{M}_{sup}=\{\bar{I}_{sup},\bar{J}_{sup}\},\ \bar{I}_{sup}=I_{sup}\cup i_o,\ \bar{J}_{sup}=J_{sup}\cup j_*.$$

The suboptimality estimate of the PSF-point $\{x,\bar{M}_{sup}\}$ is equal to

$$\beta(x,\bar{M}_{sup}) = \beta(x,M_{sup}) + \sum_{k=0}^{k_o-1}(\sigma^{k+1}-\sigma^k)\alpha^k.$$

Correcting the sets J_N^+ and J_N^- we suppose

$$\bar{J}_N^+ = \{\ j\in\bar{J}_N:\ \delta_j(\sigma^*)>0\ \}\cup\{\ j\in\{s_1,\ \ldots,s_{q-1}\ \}\cup$$

$$\cup\ j_o:\ \Delta\delta_j>0\}\cup\{j\in\{s_{q+1},\ldots,s_p\}\cap J_n^+\};$$

$$\bar{J}_N^- = \{\ j\in\bar{J}_N:\ \delta_j(\sigma^*)<0\}\cup\{\ j\in\{\ s_1,\ \ldots,\ s_{q-1}\}\cup j_o:\ \ \Delta\delta_j$$

$$<0\}\cup\{j\in\{s_{q+1},\ldots,s_p\}\cap J_N^-\}.$$

If $\beta(x,\bar{M}_{sup})\le\varepsilon$ then construction of the initial support is stopped since x is the ε-optimal feasible point. If $\beta(x,\bar{M}_{sup})>\varepsilon$, $|\bar{I}_{sup}|=rank\ A$, we delete all the linearly dependent equalities $A(i,J)x=b_i$, $i\in\bar{I}_N$, from problem (1.2). For the obtained problem x is a feasible point and $\bar{M}_{sup}$ is a support, i.e. Phase I has been finished.

In the case $\beta(x,\bar{M}_{sup})>\varepsilon$, $|\bar{I}_{sup}|<rank\ A$ we proceed to a new iteration.

6. FINITE TERMINATION OF THE ADAPTIVE METHOD.

A method for solving an extremal problem is called finite if for any initial information it reveals the unsolvability of the problem or constructs an ε-optimal feasible point after a finite number of operations. Every iteration of the adaptive method described in Sections 1-5 consists of a finite number of elementary arithmetical and logical operations. Therefore to prove the finite termination of the method it is sufficient to show the finiteness of its iteration number.

6.1. Dually non-degenerate problems.

Definition 6.1. Problem (1.2) will be called dually non-degenerate if all its accompanying dual feasible points are non-degenerate, i.e.

$$\Delta_j \neq 0, \ j \in J_N = J \backslash J_{sup}, \ \forall \ J_{sup} \subset J. \tag{6.1}$$

Every problem (1.2) has a finite number of accompanying dual feasible points since the number of its supports (such sets $J_{sup} \subset J$ that $\det A(I, J_{sup}) \neq 0$) is finite.

The relations (6.1) can be violated only for exceptional combination of the parameters c, A. Therefore for fixed dimensions m and n a measure of the set of degenerate problems is equal to zero in the space of all problems (1.2). In other words taken at random problem (1.2) will be almost certainly dually non-degenerate. If the dually degenerate problem (1.2) is solved by computer then it is interpreted as non-degenerate, owing to round-off errors.

Theorem 6.1. For $\|d_*\| < \infty$, $\|d^*\| < \infty$, $m < \infty$, $n < \infty$ and any $\varepsilon \geq 0$ the adaptive method starting with an arbitrary initial SF-point constructs an ε-optimal feasible point of every dually non-degenerate problem (1.2) in a finite number of iterations.

In the case of dual non-degeneracy the inequality $\sigma^* > 0$ holds at iterations of the adaptive method. Consequently the dual objective function strictly decreases $\Phi(\bar{\lambda}) < \Phi(\lambda)$ at

$\lambda \longrightarrow \bar{\lambda}$. Therefore there is no support which may occur twice. Since the number of supports is finite, the process of the support modification will necessarily stop. This can happen only when an ε-optimal feasible point has been constructed.

Using the same scheme one can prove a stronger statement on finiteness termination of the algorithm for the case when iterations with $\sigma^* = 0$ occur a finite number of times.

6.2. Arbitrary problems.

Finite modifications of the adaptive algorithm have been carried out in Minsk for solving dually degenerate problems. We give one of them. The modification consists in essence of a special rule for choice of j_0:

$$j_0 = min \{ j \in J_{sup} : \theta_j < 1 \}.$$

If the modification is also applied to Phase I then we obtain a finite method for solving problem (1.2) for

$$\|d_*\| < \infty, \ \|d^*\| < \infty.$$

A finite modification of the adaptive method for the case $\|d_*\| = \infty$, $\|d^*\| = \infty$ can be constructed with the help of artificial components of the simple constraints vectors:

$$d^a_{*j} > -\infty \quad \text{for} \quad d_{*j} = -\infty \quad \text{and} \quad d^{*a}_j < \infty \quad \text{for} \quad d^*_j = \infty$$

and analysis of results for $x^0_j = d^{*a}_j$ or d^a_{*j}.

REFERENCES

[1] Ackermann J. Der entwurt linearer regelungsysteme im zustandstraum. Regelungstechnik, 1972, N 20, pp.1297-300.

[2] Aström K.J. Introduction to stochastic control theory. Academic Press, New York,1969.

[3] Barbashin E.A. Introduction to stability theory. Nauka, Moscow, 1967 (in Russian).

[4] Bellman R. Dynamic programming. Princeton University Press, Princeton, 1963.

[5] Boltyanski V.G. Mathematical methods of optimal control. Nauka, Moscow, 1969 (in Russian).

[6] Bushau D.W. Experimental towing tank .Stevens Institute of Technology, Hobonen, 1953, Report 469.

[7] Callier F.M., Desoer C.A. Multivarable feedback systems. Springer-Verlag, Berlin Heidelberg New York, 1982.

[8] Chang S.S.L. Synthesis of optimum control systems. McGraw-Hill, New York Toronto London, 1961.

[9] Chernousko F.L., Kolmanovskii V.B. Optimal control under random perturbations. Nauka, Moscow, 1978 (in Russian).

[10] Cruz J.B.Jr. Feedback systems. McGraw-Hill, New York, 1971.

[11] Dantzig G.B. Linear programming and extensions. Princeton University Press, New Jersey, 1963.

[12] Eykhoff P. System identification : parameter and state estimation. Wiley and Sons Inc., London New York Sydney, 1974.

[13] Feldbaum A.A. Optimal processes in systems of automatic

regulation . Automatika and telemeh., 1953, V.14, N 6, pp.712-728.

[14] Feldbaum A.A. Fundamentals of theory of optimal automatic systems. GIFML, Moscow, 1963 (in Russian).

[15] Gabasov R., Kirillova F.M. Qualitative theory of optimal processes. Marcel Dekker Inc., New York and Basel, 1976.

[16] Gabasov R., Kirillova F.M. Methods of linear programming. Parts 1-3. BGU Publishing House, Minsk, 1977-1980 (in Russian).

[17] Gabasov R., Kirillova F.M. Constructive methods of parametric and functional optimization. In:Akashi K.(ed) Preprints of 8-th World Congress on Automatic Control, Kyoto, Japan, 1981, V.4, pp.111-116.

[18] Gabasov R.,Kirillova F.M. Consideration of optimal control problems specificity on generalizing mathematical programming algorithms. In: Gertler J., Keviczky L.(eds). Preprints of the 9-th World Congress on Automatic Control, Budapest, Hungary, 1984, V.5, pp.264-269.

[19] Gabasov R., Kirillova F.M.,et al. Constructive methods of optimization. Parts 1-4. University Press, Minsk, 1984-1991 (in Russian).

[20] Gabasov R., Kirillova F.M. New algorithms for solving extremal problems. In: Thoma M., Wyner A.(eds). Proc. of the 9-th International Conference on Analysis and Optimization of Systems, 1990, Antibes, pp.571-579.

[21] Gabasov R., Kirillova F.M., Kostyukova O.I. Dual algorithm of optimization of a linear dynamic system. PCIT, 1983, V.12, N 4, pp.253-267.

[22] Gabasov R., Kirillova F.M., Kostyukova O.I., Pokataev A.V. Optimal program controls and flexible feedback. In:

Isermann R.(ed). Preprints of 10-th World Congress on Automatic Control, 1987, Munich, FRG, V.8, pp.119-124.

[23] Gurdes A.N., Desoer C.A. Algebraic theory of linear feedback systems with full and decentralized compensators. Springer-Verlag, Berlin Heidelberg New York, 1981.

[24] Kalman R. On general theory of control systems. In: Proceedings of the 1-st IFAC congress, 1961, Academy of Sc., Moscow, V. 2, pp. 521-547 (in Russian).

[25] Knobloch H.W. Disturbance attenuation by feedback. In: Bagchi A., Jonson H.Th.(eds). Systems and optimization. Springer-Verlag, Berlin Heidelberg New York, 1984, V.66, pp.156-170.

[26] Kwakernaak H., Sivan R. Liner optimal control systems. Wiley-Interscience, New York London Sydney Toronto, 1972.

[27] Krasovskii N.N. The theory of controlled motion. Nauka, Moscow , 1968 (in Russian).

[28] Krasovskii N.N. Control by dynamic systems. Nauka, Moscow, 1977 (in Russian).

[29] Krasovskii N.N., Subbotin A.I. Positional differential games. Nauka, Moscow, 1974 (in Russian).

[30] Kurzanskii A.B. Control and observation under conditions of uncertainty. Nauka, Moscow, 1977 (in Russian).

[31] Laning J.H., Battin R.H. Random processes in automatic control. McGraw-Hill, New York, 1956.

[32] Lee E.B.,Markus L. Foundations of optimal control theory. Wiley and Sons Inc., New York London Sydney, 1967.

[33] Leondes C.T.(ed). Control and dynamic systems: advances

in theory and applications. Academic Press, New York, San Francisco and London, 1976.

[34] Lyotov A.M. Dynamics of flight and control. Nauka, Moscow, 1980 (in Russian).

[35] Lyapunov A.M. A General problem about movement stability. GITL, Moscow, 1950 (in Russian).

[36] Moroz A.I. Time-optimal feedback control for linear non-stationary systems. Int.J. Control, 1984, V.39, N 5.

[37] Moroz A.I. Course of theory of systems. Vycshaja schcola, Moscow, 1987 (in Russian).

[38] Pontryagin L.S., Boltyanski V.G., Gamkrelidze R.V. and Mischenko E. F. Mathematical theory of optimal processes. Nauka, Moscow, 1983 (in Russian).

[39.] Pugachev V.S. Theory of random functions and its application in automatic control problems. Gostechizdat, Moscow, 1957 (in Russian).

[40] Schweppe F.C. Uncertain dynamic systems. Englewood Cliffs, Prentice Hall, 1973.

[41] Thoma M. Theorie linearer regelsysteme. Vieweg, Braun-Schweig, 1973.

[42] Vidyasagar M.Control system synthesis: a factorization approach. MA: MIT Press, Cambridge, 1985.

Lecture Notes in Control and Information Sciences

Edited by M. Thoma

1992–1995 Published Titles:

Vol. 167: Rao, Ming
Integrated System for Intelligent Control
133 pp. 1992 [3-540-54913-7]

Vol. 168: Dorato, Peter; Fortuna, Luigi; Muscato, Giovanni
Robust Control for Unstructured Perturbations: An Introduction
118 pp. 1992 [3-540-54920-X]

Vol. 169: Kuntzevich, Vsevolod M.; Lychak, Michael
Guaranteed Estimates, Adaptation and Robustness in Control Systems
209 pp. 1992 [3-540-54925-0]

Vol. 170: Skowronski, Janislaw M.; Flashner, Henryk; Guttalu, Ramesh S. (Eds)
Mechanics and Control. Proceedings of the 4th Workshop on Control Mechanics, January 21-23, 1991, University of Southern California, USA
302 pp. 1992 [3-540-54954-4]

Vol. 171: Stefanidis, P.; Paplinski, A.P.; Gibbard, M.J.
Numerical Operations with Polynomial Matrices: Application to Multi-Variable Dynamic Compensator Design
206 pp. 1992 [3-540-54992-7]

Vol. 172: Tolle, H.; Ersü, E.
Neurocontrol: Learning Control Systems Inspired by Neuronal Architectures and Human Problem Solving Strategies
220 pp. 1992 [3-540-55057-7]

Vol. 173: Krabs, W.
On Moment Theory and Controllability of Non-Dimensional Vibrating Systems and Heating Processes
174 pp. 1992 [3-540-55102-6]

Vol. 174: Beulens, A.J. (Ed.)
Optimization-Based Computer-Aided Modelling and Design. Proceedings of the First Working Conference of the New IFIP TC 7.6 Working Group, The Hague, The Netherlands, 1991
268 pp. 1992 [3-540-55135-2]

Vol. 175: Rogers, E.T.A.; Owens, D.H.
Stability Analysis for Linear Repetitive Processes
197 pp. 1992 [3-540-55264-2]

Vol. 176: Rozovskii, B.L.; Sowers, R.B. (Eds)
Stochastic Partial Differential Equations and their Applications. Proceedings of IFIP WG 7.1 International Conference, June 6-8, 1991, University of North Carolina at Charlotte, USA
251 pp. 1992 [3-540-55292-8]

Vol. 177: Karatzas, I.; Ocone, D. (Eds)
Applied Stochastic Analysis. Proceedings of a US-French Workshop, Rutgers University, New Brunswick, N.J., April 29-May 2, 1991
317 pp. 1992 [3-540-55296-0]

Vol. 178: Zolésio, J.P. (Ed.)
Boundary Control and Boundary Variation. Proceedings of IFIP WG 7.2 Conference, Sophia-Antipolis, France, October 15-17, 1990
392 pp. 1992 [3-540-55351-7]

Vol. 179: Jiang, Z.H.; Schaufelberger, W.
Block Pulse Functions and Their Applications in Control Systems
237 pp. 1992 [3-540-55369-X]

Vol. 180: Kall, P. (Ed.)
System Modelling and Optimization. Proceedings of the 15th IFIP Conference, Zurich, Switzerland, September 2-6, 1991
969 pp. 1992 [3-540-55577-3]

Vol. 181: Drane, C.R.
Positioning Systems - A Unified Approach
168 pp. 1992 [3-540-55850-0]

Vol. 182: Hagenauer, J. (Ed.)
Advanced Methods for Satellite and Deep Space Communications. Proceedings of an International Seminar Organized by Deutsche Forschungsanstalt für Luft-und Raumfahrt (DLR), Bonn, Germany, September 1992
196 pp. 1992 [3-540-55851-9]

Vol. 183: Hosoe, S. (Ed.)
Robust Control. Proceesings of a Workshop held in Tokyo, Japan, June 23-24, 1991
225 pp. 1992 [3-540-55961-2]

Vol. 184: Duncan, T.E.; Pasik-Duncan, B. (Eds)
Stochastic Theory and Adaptive Control. Proceedings of a Workshop held in Lawrence, Kansas, September 26-28, 1991
500 pp. 1992 [3-540-55962-0]

Vol. 185: Curtain, R.F. (Ed.); Bensoussan, A.; Lions, J.L.(Honorary Eds)
Analysis and Optimization of Systems: State and Frequency Domain Approaches for Infinite-Dimensional Systems. Proceedings of the 10th International Conference, Sophia-Antipolis, France, June 9-12, 1992.
648 pp. 1993 [3-540-56155-2]

Vol. 186: Sreenath, N.
Systems Representation of Global Climate Change Models. Foundation for a Systems Science Approach.
288 pp. 1993 [3-540-19824-5]

Vol. 187: Morecki, A.; Bianchi, G.; Jaworeck, K. (Eds)
RoManSy 9: Proceedings of the Ninth CISM-IFToMM Symposium on Theory and Practice of Robots and Manipulators.
476 pp. 1993 [3-540-19834-2]

Vol. 188: Naidu, D. Subbaram
Aeroassisted Orbital Transfer: Guidance and Control Strategies
192 pp. 1993 [3-540-19819-9]

Vol. 189: Ilchmann, A.
Non-Identifier-Based High-Gain Adaptive Control
220 pp. 1993 [3-540-19845-8]

Vol. 190: Chatila, R.; Hirzinger, G. (Eds)
Experimental Robotics II: The 2nd International Symposium, Toulouse, France, June 25-27 1991
580 pp. 1993 [3-540-19851-2]

Vol. 191: Blondel, V.
Simultaneous Stabilization of Linear Systems
212 pp. 1993 [3-540-19862-8]

Vol. 192: Smith, R.S.; Dahleh, M. (Eds)
The Modeling of Uncertainty in Control Systems
412 pp. 1993 [3-540-19870-9]

Vol. 193: Zinober, A.S.I. (Ed.)
Variable Structure and Lyapunov Control
428 pp. 1993 [3-540-19869-5]

Vol. 194: Cao, Xi-Ren
Realization Probabilities: The Dynamics of Queuing Systems
336 pp. 1993 [3-540-19872-5]

Vol. 195: Liu, D.; Michel, A.N.
Dynamical Systems with Saturation Nonlinearities: Analysis and Design
212 pp. 1994 [3-540-19888-1]

Vol. 196: Battilotti, S.
Noninteracting Control with Stability for Nonlinear Systems
196 pp. 1994 [3-540-19891-1]

Vol. 197: Henry, J.; Yvon, J.P. (Eds)
System Modelling and Optimization
975 pp approx. 1994 [3-540-19893-8]

Vol. 198: Winter, H.; Nüßer, H.-G. (Eds)
Advanced Technologies for Air Traffic Flow Management
225 pp approx. 1994 [3-540-19895-4]

Vol. 199: Cohen, G.; Quadrat, J.-P. (Eds)
11th International Conference on Analysis and Optimization of Systems – Discrete Event Systems: Sophia-Antipolis, June 15–16–17, 1994
648 pp. 1994 [3-540-19896-2]

Vol. 200: Yoshikawa, T.; Miyazaki, F. (Eds)
Experimental Robotics III: The 3rd International Symposium, Kyoto, Japan, October 28-30, 1993
624 pp. 1994 [3-540-19905-5]

Vol. 201: Kogan, J.
Robust Stability and Convexity
192 pp. 1994 [3-540-19919-5]

Vol. 202: Francis, B.A.; Tannenbaum, A.R. (Eds)
Feedback Control, Nonlinear Systems, and Complexity
288 pp. 1995 [3-540-19943-8]

Vol. 203: Popkov, Y.S.
Macrosystems Theory and its Applications: Equilibrium Models
344 pp. 1995 [3-540-19955-1]

Vol. 204: Takahashi, S.; Takahara, Y.
Logical Approach to Systems Theory
192 pp. 1995 [3-540-19956-X]

Vol. 205: Kotta, U.
Inversion Method in the Discrete-time Nonlinear Control Systems Synthesis Problems
168 pp. 1995 [3-540-19966-7]

Vol. 206: Aganovic, Z.; Gajic, Z.
Linear Optimal Control of Bilinear Systems
148 pp. 1995 [3-540-19976-4]